THE SEARCH FOR
EXTRATERRESTRIAL
INTELLIGENCE

Are we alone in the universe? Is the search for extraterrestrial intelligence a waste of resources or a genuine contribution to scientific research? And how should we communicate with other life-forms if we make contact?

The search for extraterrestrial intelligence (SETI) has been given fresh impetus in recent years following developments in space science which go beyond speculation. The evidence that many stars are accompanied by planets; the detection of organic material in the circumstellar disks of which planets are created; and claims regarding microfossils on Martian meteorites have all led to many new empirical searches.

Against the background of these dramatic new developments in science, *The Search for Extraterrestrial Intelligence: a philosophical inquiry* critically evaluates claims concerning the status of SETI as a genuine scientific research programme and examines the attempts to establish contact with other intelligent life-forms of the past thirty years. David Lamb also assesses competing theories on the origin of life on Earth, discoveries of ex-solar planets and proposals for space colonies as well as the technical and ethical issues bound up with them. Most importantly, he considers the benefits and drawbacks of communication with new life-forms: how we should communicate and whether we could.

The Search for Extraterrestrial Intelligence is an important contribution to a field which until now has not been critically examined by philosophers. David Lamb argues that current searches should continue and that space exploration and SETI are essential aspects of the transformative nature of science.

David Lamb is honorary Reader in Philosophy and Bioethics at the University of Birmingham.

THE SEARCH FOR EXTRATERRESTRIAL INTELLIGENCE

A philosophical inquiry

David Lamb

London and New York

First published 2001
by Routledge
2 Park Square, Milton Park, Abingdon, Oxon, OX14 4RN

Simultaneously published in the USA and Canada
by Routledge
29 West 35th Street, New York, NY 10001

Routledge is an imprint of the Taylor & Francis Group

Transferred to Digital Printing 2004

© 2001 David Lamb

Typeset in Baskerville by Taylor & Francis Books Ltd

British Library Cataloguing in Publication Data
A catalogue record for this book is available from the British Library

Library of Congress Cataloging in Publication Data
Lamb, David, 1942–
The search for extraterrestrial intelligence : a philosophical inquiry /
David Lamb.
Includes bibliographical references and index.
1. Life on other planets.   I. Title.
QB54 .L35 2001
999–dc21
00-062787

ISBN 0–415–24341–6 (hbk)
ISBN 0–415–24342–4 (pbk)

FOR ESMOND

CONTENTS

ACKNOWLEDGEMENTS

I would like to thank a number of people for their valuable assistance during the preparation of this book. Dr Doris Schroeder for finding those elusive sources which often turned out to be of immense value, and Dr Susan Easton who patiently read through at least two versions of the manuscript. My interest in SETI as a scientific inquiry emerged in the course of teaching the philosophy of science to university students during the 1990s. An earlier account of some of my thoughts on this subject can be found in a paper entitled 'Communication with extraterrestrial intelligence: SETI and scientific methodology', first published in D. Ginev and R.S. Cohen (eds) *Issues and Images in the Philosophy of Science* (1997: 223–51), a revised version of which is republished here with kind permission from Kluwer Academic Publishers. Special thanks to Richard Scothern, who shares my interest in the intelligence of non-human terrestrial beings, and who not only supplied me with valuable computer hardware but managed to bring me to a reasonable level of computer literacy. Thanks to my very good friend, Gonzalo Munévar, for his conversations, wisdom and wit, which I have been fortunate to enjoy over many years. Thank you also to the copy editor, Susan Dunsmore. Needless to say, I claim responsibility for any remaining errors.

ABBREVIATIONS

ASEPS	astronomical studies of extrasolar planetary systems
BUFORA	British Unidentified Flying Object Research Association
CAT	consciousness-assisted technology
CE	close encounter
CETI	communication with extraterrestrial intelligence
DUMAND	Deep Underwater Muon and Neutrino Detector
EBEs	extraterrestrial biological entities
EEG	electroencephalogram
ESA	European Space Agency
ETA	extraterrestrial activity
ETI	extraterrestrial intelligence
FRESIP	frequency of Earth-sized planets
FTL	faster-than-light
GAO	Government Accounting Office
GHz	gigahertz
GRBs	Gamma-ray bursts
HRMS	High Resolution Microwave Survey
IFO	identified flying object
IRAS	infrared astronomical satellite
LGM1	little green man 1
MCSA	Multi-Channel Spectrum Analysers
META	Megachannel Extraterrestrial Assay
MOP	Microwave Observing Project
NASA	National Aeronautical and Space Administration
NRAO	National Radio Astronomy Observatory
OBEs	out of body experiences
RFI	radio frequency interference
SCs	supercivilizations
SEP	solar electric propulsion
SETHI	search for extraterrestrial human intelligence
SETI	search for extraterrestrial intelligence
SF	science fiction

ABBREVIATIONS

SLIME	subsurface lithoautotrophic microbial ecosystem
TAC	technology-assisted consciousness
TOPS	Towards Other Planetary Systems
UFO	unidentified flying object
USAAF	United States of America Air Force
UV	ultraviolet
VR	virtual reality

INTRODUCTION

I believe this nation should commit itself to achieving the goal, before this decade is out, of landing a man on the Moon and returning him safely to Earth. No single space project in this period will be more impressive to mankind or more important for the long-range exploration of space; but none will be so difficult or expensive to accomplish.

(President John F. Kennedy, message to Congress, 25 May 1961)

Despite a wealth of science fiction and scientific speculation on the origins of the universe and life itself, professional philosophers have shown little interest in space exploration. Our culture is very much an Earthbound one and provides little imaginative foothold for the exploration of space. Yet we stand on the threshold of space exploration. We may decide to refrain from further explorations and draw to a halt the expansion of our technological culture, or we may begin a project of exploration that will, one day in the far-off future, lead to the stars. For most of us, while space travel is scientific fact, communication with extraterrestrials is science fiction (SF). But the line between the two is not rigid. Why should we, it might be asked, be the only culture in the universe to contemplate exploration? This question leads directly to the subject matter of this book, the investigation of theories and beliefs concerning the existence of intelligent extraterrestrial forms of life and the proposals for making contact with them. The name given to scientific research into the search for extraterrestrial intelligent life is SETI, that is, the Search for Extraterrestrial Intelligence. One of the objectives of writing this book is to evaluate the respective claims that have been made about the status of SETI as a serious scientific enterprise and the variety of hypotheses which have emerged during the past three or four decades of SETI research.

How does SETI research differ from science fiction? Both concern life on other worlds, conquest of space, communication with advance cultures and the possibility of a galactic community. There are many who would argue that science and SF are interchangeable, and that what is SF today could be routine science tomorrow. In this context SF prophets, such as Jules Verne and H.G. Wells are often cited as visionaries who imagined what scientists would later

1

discover. But this obscures a very important distinction between the scientific imagination and the literary imagination. Whereas SF, like an entertainment magician, requires a suspension of our demand for an explanation, science even at its most speculative and imaginative, does not ignore or try to conceal the improbable. SF requires a massive leap of imagination, skilfully concealing the gaps in our knowledge; the better the tale, the better the gaps are concealed and the awkward questions remain unasked. This is partly why there is a mass market for UFO theories and accounts of alien abductions; good tales have the ability to drive out boring scientific truths. But SETI must face the requirement to present details and explanations at every stage. Of course, when reasoning in the context of discovery, not every claim has to be secured with hard evidence and impregnable theory, but the scientist must be willing to try to provide explanations, not conceal things. SETI investigation is played according to the rules of science. Its speculations are – or at least ought to be – restricted to known areas of science. This is why its exponents pursue radio searches, not spirit guides or super-rockets. SETI's imaginative extensions of scientific thinking are not arbi-trary, but even when non-quantifiable, they are subject to the requirement to provide proof, coherence and justification.

An important rule, which is frequently applied in SETI, is a prohibition on attempts to attribute skills and technology to extraterrestrials which cannot be demonstrated on Earth. This rule is applied, for example, when responding to claims that extraterrestrials are visiting Earth with the assistance of energy sources we cannot comprehend. Or, to consider a more recent example, the bulk of SETI research is based on a search of radio transmissions which assume that extraterrestrials need possess nothing beyond the level of communication tech-nology on Earth. However, recent developments in laser technology, and improvements in the design of equipment to detect potential laser signals, support the view that an optical search is worthy of consideration. If we on Earth have mastered these technologies, then it is plausible to assume that they are not beyond the reach of another scientific culture somewhere in the universe. Now they may – if they exist at all – have developed all kinds of exotic methods for communicating, but a scientific approach demands that we should only search for something similar to our current level of technology. Neverthe-less, despite a commitment to current scientific achievements, SETI operates at a level comparable to the context of discovery in scientific research, where rules are more loosely applied, and where the short-term objective is the plausibility of its hypotheses not their certitude. It is important to stress the distinction between the context of discovery and the context of justification. While it is maintained here that both involve rational processes, the former context is where analogies might be pursued, where hypotheses are generated, enter-tained, considered or ruled out. Although reasoning in the context of discovery is rule-governed (Lamb, 1991), the requirement for rigorous proof and evidential support, which is a characteristic of the context of justification, may be suspended.

INTRODUCTION

There are, nevertheless, limits to arguments and speculation in the context of discovery. For example, one cannot dismiss objections to one's speculations by simply retorting that in the future, with technological advances, all problems can be solved. But we can speculate on solutions to problems on which we have made a convincing start. SETI might well turn out to be an example of too much speculation chasing too few facts, but this is not an argument for dismissing its conjectures as nonsense; rather, it indicates a requirement to go ahead and produce more facts.

It is important, therefore, to maintain a distinction, which is actually upheld by many SETI researchers, between the conceivably possible and the probable. Conjectures about advanced extraterrestrial civilizations are conceivably possible, but as we have no evidence of them, they are not candidates for probabilistic argument. Nevertheless, these conjectures are not unscientific; they may be essential to the scientific project under consideration and contribute to modes of reasoning in the context of discovery, which have been frequently ignored by philosophers who have confined their attention to modes of reasoning in the contexts of justification. It must be stressed that scientific speculation about possible sites where life could exist, and speculation about methods of communication with other forms of intelligence, do not belong in finished research reports which can be analysed by philosophers and historians of science with an interest in the context of justification. Rather, these speculations belong in the context of discovery, where ideas are generated, analogies are pursued, and current theory is extended. This, of course, is part of the rational process of scientific development.

Several dramatic developments have intensified interest in SETI: first, the claim that many stars are accompanied by planets; and second, detection of organic material in the circumstellar disks from which planets are created. Both these claims suggest that there are numerous potential sites for life. During the past few years there have also been many improvements in experimental methods for the detection of life, and further improvements in the technological capacity to conduct scientific searches for intelligent radio and optical signals. All of this was given a significant boost when in August 1996 NASA (National Aeronautical and Space Administration) scientists claimed that life may once have existed on Mars. The suggestion that life could independently appear on two planets within the same stellar system generated considerable speculation regarding the extent of extraterrestrial intelligence. While claims regarding ex-solar planets have been widely accepted since 1995, speculation regarding former life on Mars has met with sceptical opposition. But if evidence concerning ex-solar life is confirmed, the consequences are profound: beliefs associated with the alleged uniqueness of human life and intelligence will be challenged.

The proposals for space exploration and possible contact with extraterrestrial life, which are discussed in this book, have emerged in the context of scientific debate in some of the leading scientific journals. The history of SETI and its relationship with various branches of science has been documented in several major publications (Crowe, 1988; Dick, 1996, 1998). The approach taken in this book is not so much a historical survey or a study of works of science fiction but an examination of a variety of proposals and speculations that have emerged in the work of prominent scientists, some of whom are leading figures in their respective disciplines. Chapter 1 examines various proposals to communicate with forms of extraterrestrial intelligence (ETI), from Kant's nineteenth-century successors to the most recent searches by radio astronomers throughout the world. Chapter 2 evaluates the status of SETI as a branch of scientific inquiry and considers some of the methodological objections to it. The latter part of Chapter 2 focuses on the Drake equation which, so it is claimed, provides SETI research with a theoretical framework. It is argued that SETI is both a branch of scientific inquiry and a campaign for exploration.

Chapter 3 develops some of the issues raised in Chapter 2's discussion of the Drake equation, namely the search for habitable planets, requirements for the development of intelligent life in the universe, theories on the origin of life, and the possibility of exotic forms of extraterrestrial life. Chapter 4 evaluates the search for life in the solar system, proposals for terraforming potentially habitable sites and proposals for space colonization. Any speculation about space travel raises questions about possible extraterrestrial visits to Earth. This topic is covered in Chapter 5, which conducts an investigation into the status of Ufology and the claims that extraterrestrials have attempted to visit Earth. Chapter 6 assesses some of the arguments that have been employed against SETI's assertion that the galaxy may be populated with many forms of ETI. One of the foremost assertions of the uniqueness of human intelligence is derived from the fact that no authentic contact has yet been made. Arguments in this vein frequently invoke what is known as 'Fermi's Paradox', named after the Italian physicist, Enrico Fermi, who responded to arguments in favour of a galaxy populated with ETI with the retort 'Where are they?' While one authenticated contact would resolve this problem there is, nevertheless, an intellectual requirement that SETI researchers respond to the assertion that so far the universe has been silent. Chapter 6 examines SETI's attempt to explain the 'great silence' and resolve Fermi's Paradox. Chapter 7 casts a critical eye over the claim that super-advanced civilizations – perhaps a billion years ahead of us – may exist somewhere in the universe. Questions are also raised with regard to the proposed benefits of contact with an advanced civilization.

Philosophy and theology have always been concerned with speculation on the 'why' of human existence and its alleged uniqueness. Technological developments during the past twenty years have placed us in a position to embark on an observational attack on these questions. With space probes and new observation technologies the question of whether there is ex-solar life can now be answered with scientific rigour. For the first time, it will be possible, as we enter the twenty-first century, for science and philosophy to join forces in addressing one of the most fundamental questions concerning our existence.

1

COMMUNICATION WITH EXTRATERRESTRIAL INTELLIGENCE

Gort, Klaatu berada nikto.

(The Day the Earth Stood Still, Warner Brothers, 1951)

Introduction

This chapter examines proposals for communication with extraterrestrial intelligence (ETI). Some of the ancient cosmologies, such as the Aristotelian model of the universe, are described by way of an introduction to the post-Copernican vision of a universe populated with many worlds similar to our own. Nevertheless, it will be argued that pre-Copernican beliefs regarding the centrality of the Earth in the cosmos and analogies with terrestrial conditions have survived in modern astronomy. Early in the twentieth century, beliefs in the existence of Martians drew heavily on scientific and political analogies with life on Earth. Modern radio searches, which offer the first empirical solution to the search for ETI, are nevertheless dependent upon analogies with technological developments during the past century on Earth. This chapter ends with some brief speculations regarding the effects of contact with ETI and the possibility of a galactic network of communicating beings.

Ancient cosmology and the Copernican revolution

The desire to make sense of the cosmos and understand our place in it is as old as human history. From earliest times people have speculated on the nature of the universe and have gazed upon and wondered at the night sky. Out of this wonderment emerged the earliest cosmologies and beliefs about the origins of earthly life and the possibility of extraterrestrial life. The ancient cosmologies, however, had one feature in common; they were fashioned on terrestrial concerns. Thus the heavens were perceived of as an enclosure for the Earth, and were peopled with beings who were not fundamentally distinct from those found on Earth. Indeed, the Gods and other extraterrestrial beings were frequently portrayed in various forms of intercourse with the inhabitants of Earth. The

Egyptians, for example, depicted a heaven consisting of platters surrounded by water, ultimately supported by a God. This perception was a reasonable reflection of the Nile environment. And just as human nobles and kings rode in horse-drawn chariots so too did the various Gods make their way across the night sky in heavenly chariots.

One of the influential Greek cosmologists was Anaximander of Miletos, who depicted the stars as portions of air in the shape of rotating wheels filled with fire. The Sun was a chariot wheel 28 times the size of the Earth, and was also made of fire. This indicated an advance over the Egyptians, as Anaximander described heavenly bodies as natural mechanisms, rather than Gods. From the fourth century BC the Greeks saw the Earth as a tiny sphere suspended stationary at the geometric centre of a larger rotating sphere which carried the stars. Beyond that – nothing, no space, no matter. In fact, this is the view that the modern world inherited from the ancient world. Thomas Kuhn (1957: 27) described it as 'the two-sphere model of the universe'. This framework of thought found expression in the idea of an interior sphere for human beings and Earth-life, and an external sphere for the stars and heavenly forms of (after) life. It was a model which was to survive, with variation, for about 2,000 years. It was supported by observations of the 'domed' heavens of the Egyptians and Greeks, and by aesthetic opinion related to the apparent uniform rate of motion of the stars. Plato's *Timaeus* is a classical portrayal of the two-sphere model which is based on an appeal to aesthetic and religious grounds.

Appeals to conceptual economy also lent support to the two-sphere model. The heavens are not displayed in a random pattern, but reveal an order which can be comprehended. The most favoured model of the universe was of a gigantic sphere, bearing the stars, rotating westwards on a fixed axis every 23 hours and 56 minutes, with the Sun moving steadily eastwards around the elliptic once every 365 and a quarter days. With this model one can organize observations of the heavens, apply them to the practical art of navigation, and employ them for astrological predictions. It might be noted that astrology, which depicts human destiny in terms of the positions of the stars, would naturally favour a two-sphere model of the universe.

The foremost exponent of the two-sphere model was Aristotle, who held that the cosmos was closed and finite, and consisted of a series of revolving spheres with an immobile Earth at the centre. The natural motion of heavenly bodies, he held, was circular and the celestial spheres circled the Earth while the Sun, Moon and planets moved in complex circles within these spheres. About 150 AD Aristotle's cosmology was developed by Claudius Ptolemy who constructed an astronomical table, the *Almagest*, which provided a means of predicting the positions of the planets. The *Almagest* was also employed for practical purposes such as navigation, the prediction of eclipses and dates of the equinoxes.

Despite numerous modifications, the Aristotelian model of the universe remained dominant until well into the sixteenth century and was supported by astrology, astronomy, theology and common-sense experience. Unfortunately the *Algmagest* never provided a wholly satisfactory method of predicting planetary motion and as improvements to observational techniques were made, these inaccuracies provided increasing problems for Ptolemaic astronomy. During the first decade of the sixteenth century Nicolaus Copernicus, a distinguished mathe-matician and astronomer, attempted to revise Ptolemy's *Almagest*, but he came to the conclusion that the Sun had to be located at the centre of the cosmos, and that the Earth, Moon and the planets all circle the Sun. The implications of this revision were enormous: the Earth no longer occupied the central position in the universe and was no longer held to be immobile. Furthermore, as merely one more planet in motion around the Sun, the Earth no longer occupied a unique and privileged position in the cosmos.

Today we speak of the Copernican revolution, but it must be emphasized that this is not how it appeared to Copernicus himself. Copernicus had been educated in the scholastic tradition and sought to show how his theory was supported by ancient authorities. Thus his *De Revolutionibus* was not so much a revolutionary text as a revolution-making text. For in the hands of his successors Ptolemaic astronomy and the Aristotelian cosmology were overturned. Moreover, Copernicus did not resolve all the anomalies of Ptolemaic astronomy, but he did convince Galileo and others that a Sun-centred astronomy was the direction to follow.

The post-Copernican cosmology took away most of the Aristotelian theories of interplanetary matter and motion. A new generation of astronomers and cosmologists went on to reject the two-sphere model, introduce the notion of infinite space, and even dreamt of worlds beyond our solar system and the possibility of extraterrestrial life. After Copernicus published his Sun-centred account of the planetary system, others were led into speculations regarding extraterrestrial life. With the aid of his telescope Galileo not only demolished the idea of an Earth-centred universe, but gave credibility to questions concerning the very uniqueness of Earth as an abode of life.

Attempts to establish contact with extraterrestrial intelligence have a long history. People have always believed in extraterrestrial beings; they called them Gods and named the stars and planets after them. In fact, the majority of pluralist theories have been associated with religious and mystical beliefs. In scientific terms, most of the early pluralist theories did not have 'good reasons' to support them. The process of bestowing scientific credibility on pluralistic theories of life in the universe nevertheless matured slowly with Copernicanism. However, the pre-Copernican belief that the Earth was the centre of creation survived in a religious and psychological sense long after the sixteenth century, and terrestrial interests continued to dominate speculation about other forms of -life. Terrestrial analogies were employed in early telescopic observations of the Moon and Mars, which led to the naming of oceans and seas, thus influencing the expectation of other similarities. Most efforts to establish contact have revealed assumptions that ETI, should it exist, would be essentially similar to human intelligence, but perhaps a little more advanced. This chapter provides a brief survey of beliefs in ETI and of attempts to establish contact with our galactic neighbours.

The search for intelligent life in the universe

If extraterrestrial life is detected by one of the space probes or radio searches, then ours may be the last generation to think of itself as alone in the universe. If this is the case, then our successors will have a yardstick to ridicule us with, just as we now ridicule those who insisted that the Earth was flat. Speculation concerning extraterrestrial life is as old as philosophy itself. Aristotle maintained that the Moon was inhabited and, despite the collapse of the Aristotelian world-view in the sixteenth and seventeenth centuries, many supporters of the Copernican revolution were to revise Aristotle's pluralism. When Galileo expressed puzzlement over the lunar cavities which he had detected with his telescope, Kepler offered the explanation that they were formed by intelligent inhabitants who had made their homes in caves. The power of terrestrial analogies influenced early telescopic observations by Galileo's contemporaries. Kepler and others saw evidence of water, clouds and rain in the dark patches on the Moon. Thus Kepler, in what might be seen as the first work of science fiction, wrote of lunar inhabitants:

> [who] have no fixed abode, no established domicile. In the course of one of their days, they roam in crowds over their whole sphere, each according to his own nature: some use their legs, which far surpass those of our camels; some resort to wings; and some follow the receding water in boats; or if a delay of several more days is necessary, then they crawl into caves.
>
> (1967: 27)

The Copernican revolution, which eroded belief in the uniqueness of Earth, fostered a range of pluralist speculations which were partly responsible for the Church's hostility to Galileo. The pluralist beliefs of Giordano Bruno played an important part – perhaps not as decisive as nineteenth-century historians held – in his burning by the Inquisition on 16 February 1600. Among the problems which worried the Church was the suggestion that the human race might not have descended from Adam, and whether Christ would have to be crucified many times in many worlds.

In his comprehensive survey of the debate on pluralism, Michael J. Crowe (1988) introduces the young Kant as one of the forerunners of SETI. In 1755 Kant published his *Universal Natural History and Theory of the Heavens* in which he assessed the intelligence of beings throughout the universe. According to Kant, the cosmos was characterized by chaos at the centre and increasing order towards the remoter regions, such that 'the most perfect classes of rational beings

[are] farther from the centre than near it' (Kant, 1981: 167). This hierarchy of being stretched from the dullards to the near Divine. Crowe (1988: 52) draws attention to two assumptions underpinning Kant's pluralism: '[1] the great chain of being, and [2] the idea that from the type of matter dominant on a planet, one can make inferences about its inhabitants'. The first assumption is seen in Kant's comment that: 'From the highest class of thinking beings to the most abject insect, no member is indifferent to nature; and that nothing can be missing without breaking up the beauty of the whole, which consists in interconnectedness' (Kant, 1981: 185). The second assumption is found in Kant's formulation of a rule according to which intelligence is assessed in relation to the position occupied by living beings in the universe. As Kant says: 'the whole range of their perfection stands under a certain rule, according to which these become more excellent and perfect in proportion to the distance of their habitats from the Sun' (ibid.: 189). From this perspective Kant's Mercurians and Venusians are dullards. We on Earth, however, occupy the 'middle rung', but the Jovians and the Saturnians are greatly superior beings. He states: 'From one side we saw thinking creatures among whom a man from Greenland or a Hottentot would be a Newton, and on the other side some others would admire him as [if he were] an ape' (ibid.: 190). Anticipating Carl Sagan's immortals, Kant suggested that the inhabitants of the outer planets were free from decay and death, and speculated on whether those who were fortunate to live on the larger planets were 'too noble and wise' to sin. It might be interesting to consider what Kant would have said about the inhabitants of the outer planets, Uranus, Neptune and Pluto, discovered some time later. It is comforting to note that for Kant the Martians, at least, suffered from the same sinful afflictions as Earth beings.

Later in life Kant was to achieve fame for his critique of speculative systems and consequently prohibited republication of the chapters of his book that dealt explicitly with ETI. He never proposed any means of verifying his beliefs in ETI, although his faith in pluralism persisted in his mature works. For example, in the *Critique of Pure Reason* of 1781, when comparing opinion, knowledge and belief, Kant assesses the strength of various beliefs, some of which can be measured by the amount one is prepared to support with a bet. There are, he says, analogons in purely theoretical circumstances, which he calls doctrinal beliefs. Thus:

> I should not hesitate to stake my all on the truth of the proposition – if there were any possibility of bringing it to the test of experience – that, at least, some one of the planets which we see is inhabited. Hence I say that I have not merely the opinion, but the strong belief, on the correctness of which I would stake many of the advantages of life, that there are inhabitants of other worlds.
>
> (Kant, 1956: 468)

The Lunarians make a brief appearance in the Transcendent Dialectic, as evidence that:

> the objects of experience ... are not things in themselves, but given only in experience, having no existence apart from and independently of experience. That there may be inhabitants in the Moon, although no one has ever observed them, must certainly be admitted; but this assertion means only, that we may in the possible progress of experience discover them at some future time.
>
> (ibid.: 297)

Kant did not offer any guidelines on how experience could progress towards contact with the Lunarians but others were to follow him with more practical suggestions. In the 1820s the German mathematician, Carl Friedrich Gauss, devised a method of communicating the level of Earth intelligence to the Lunarians. One scheme involves mirrors which would reflect the Sun's light back to the Lunarians. He also proposed that groves of pine trees should be planted in Siberia, laid out in the shape of squares on the sides of a right triangle. Inside the triangle wheat was to be grown to contrast with the dark green of the trees. Gauss anticipated that the Lunarians, peering through their telescopes, would be able to recognize that the patterns of the pine trees were intelligently designed and, by calculating that the sum of the square of the number of pine trees forming the legs of the right triangle was equal to the number of trees in the square of the hypotenuse, would be able to infer that human intelligence had mastered Pythagoras's Theorem. A similar scheme was proposed by Joseph von Littrov, Director of the Vienna Observatory in 1840, who envisaged a 20-mile ditch in the Sahara Desert filled with kerosene and then set alight to function as a beacon. Neither scheme was implemented.

In 1941 Sir James Jeans suggested that searchlights be used to signal prime numbers to the Martians during the planet's close approach to Earth. Other proposals to communicate with Martians have included the building of large mirrors to reflect flashing semaphore messages and similar arrangements of large identifiable black sheets that would flap rhythmically.

On 16 November 1974 Frank Drake sent a 3-minute message from the 305-metre radio telescope in Arecibo, which gave details of our position in the universe, our mathematics and several physical details of human terrestrials. The message was sent to M-13, the Great Cluster in the constellation of Hercules. This message did not elicit a reply, but it provoked a strong rebuke from the Astronomer Royal, Sir Martin Ryle, in England, who argued that humanity had been exposed to threats from ETs with evil intention. This was a rather pointless protest as ordinary radio broadcasts are capable of giving away our location provided that they are 1 per cent stronger than the radio noise from the Sun. Each year the spreading waves of radio signals reach a further twenty stars.

Intelligent life on Mars

The most popular location of extraterrestrial life has been Mars, the 'red planet'. It has a number of similarities with Earth: it revolves on its own axis every 24 hours, give or take a few minutes, just like the Earth. Shapes were identified by nineteenth-century astronomers which resembled continents, polar caps and seas. In the late nineteenth century the meticulous Italian astronomer, Giovanni Schiaparelli, director of the Milan Observatory, discovered a complicated network of dark lines, which were interpreted as canals. In fact, Schiaparelli employed the Italian word *canali* which can be interpreted as channel, which is suggestive of natural causes, but it can also be interpreted as 'canal' which is suggestive of intelligent design. Some of his contemporaries chose the latter interpretation. In 1892 he observed an apparent doubling effect on several of the lines which many of his contemporaries interpreted as evidence of massive engineering works, and so began the 30-year-long saga of the Martian canals, which raged until Eugene Antoniadi, working with a more powerful 33-inch refractor telescope at Meudon Observatory, established that the 'canals' were an optical illusion.

In the first decade of the twentieth century the 'discovery' of men on Mars by Percival Lowell, the astronomer, was portrayed in his best-selling books, *Mars and its Canals* (1906) and *Mars as the Abode of Life* (1909). In these books Lowell outlined details of the Martians' canal system and their agricultural methods. Some 437 canals, ranging in length from 250 miles to 3,450 miles were reported to be linked with populated oases. Subsequently, popular fiction books by Edgar Rice Burroughs, which sold in their millions, encouraged expectations of Martian life. In 1904 a planet-wide dust storm was observed and it produced a shape identified as the letter W, which gave rise to speculation that the Martians were preparing for war. Science and fiction (Lowell and Rice Burroughs) had thus united to produce a climate in which the public mind was prepared for Martian encounters. During the twenty or so years of the Martian canal controversy, scientific opinion covered an entire range of views from those who endorsed Lowell's theory of intelligent construction to those who rejected the reality of the canals. The saga of the Martian canals is an interesting example where indefinite scientific results invite a broad role for speculation and preconception. An excellent survey of the canals' controversy and the process of scientific reasoning from observation to theory is found in Steven J. Dick (1996).

We now have a different truth. The 'canals' were optical illusions partly fostered by the fertile imaginations of astronomers. Explorations of Mars have been underway for over thirty years, beginning with the Mariner 4 spacecraft which flew past in 1965 taking twenty-one pictures of 1 per cent of the surface. Mariner 4 dashed hopes of revealing Mars as Earth's twin when it detected neither canals nor habitation. Space probes have, however, revealed great natural canyons, one of which is 2,500 miles long and 150 miles wide. The 'seas' are the result of dark basalt rock swept by violent winds, but there are said to be dry

11

beds of ancient rivers. During the 1970s the Viking spacecraft found the planet rich in oxygenated compounds, but found no signs of primitive biological life immediately below the surface. Mars is extremely cold and inhospitable. Its average daily temperature is around -60°C; its atmosphere is almost entirely composed of carbon dioxide, and there are winds of up to 450 kph with vast global dust storms. Nevertheless, Mars still manages to generate scientific controversy. The current debate over alleged fossils of Martian micro-organisms is a long way from claims of intelligently designed canals, but it still captures the public imagination.

The popular image of ETI

Popular interest in ET contact is strong, as seen in the reception of films like *Independence Day*, *Close Encounters* and *ET: The Extraterrestrial* (three of the most commercially successful films ever made), and the success of the TV series, *The X-Files*, as well as thousands of books on the subject and tabloid attention to even the wildest accounts of ET encounters. ET fever has been known to grip whole towns. In 1988 Tom Weber founded a group in Wisconsin called UFO Site Center Corp, with the objective of building a landing site for ET spacecraft. The American journalist, Howard Blum (1990), tells the story of Tom Weber's attempt to raise $50 million for the landing site outside a small US town. To attract the ETs it was considered necessary to communicate the peaceful intentions of the locals by means of a larger-than-life illuminated picture of a human greeting an alien in a spirit of friendship. No aliens were encountered but the townspeople were exposed to a familiar philosophical problem. How does one depict friendship to ETs? One suggestion, which was rejected, involved a large model of a man and a woman copulating. Those who supported this suggestion argued that the primal scene depicted peaceful intentions. Eventually they chose a handshake. The human, tall, slim and Aryan, and the alien with two solid feet and a head shaped like a watermelon. Blum reports from a meeting of the townspeople:

> Still, there was still some concern after a mock-up of the pair of figures at one Wednesday meeting. Is it possible, someone asked, that a handshake might not mean the same in the Andromeda Galaxy as it does in Chippewa Falls? Suppose, it was suggested, a handshake is a vulgar gesture to an alien? That we're illuminating a cosmic 'screw you' to the first visitors from another world? Or perhaps a handshake could mean 'Let's fight'? Or even 'Good Bye'? But Tom Weber was undeterred. 'If they're smart enough to get here, they're smart enough to figure everything else out', he ruled. The handshake would remain.
>
> (Blum, 1990: 185)

Such examples of anthropocentrism are as amusing as they are obvious. But are the SETI scientists and the bioastronomical community immune from it and equally as selective in their approach?

Attempts to communicate with ETI during the 1970s involved the Pioneer and Voyager spacecrafts. Pioneers 10 and 11 were launched in 1972 towards Jupiter and Saturn on trajectories that would eventually take them beyond the solar system. They carried plaques depicting men and women. The background story of these plaques reveals much about terrestrial politics of the late twentieth century. The drawing of the man and woman depicted them without any clothes. While no genitalia were included in the drawing, NASA was strongly condemned in the US media for sending pornography into space. The faces of the human figures had to be modified to be more ethnically diverse, and women's representatives questioned why only the man's hand was raised in a peaceful greeting. In the summer of 1977 Voyagers I and II were launched by NASA from Cape Canaveral on an exploratory mission to return data from Jupiter, Saturn, Uranus and Neptune, then leaving the solar system for an endless journey through space. These spacecraft carry a 12-inch copper disc entitled *Sounds of the Earth*, which carries a message from Kurt Waldheim, former Secretary General of the United Nations, and a message from the US President, James Carter, saying that the USA is a 'community of 240 million human beings among the more than 4 billion who inhabit the Earth'. The disc also contains a selection of sounds of footsteps, heartbeats, laughter and a sample of music from Stravinsky, Louis Armstrong and Chuck Berry. The technology which would be used to play the message is already obsolete on Earth. Were Voyager to return in a couple of centuries it would be extremely difficult to extract its message. Other information aboard the craft includes an essay on the evolution of the Earth, our location in the galaxy, pictures of a forest scene, snowflakes, a supermarket, an elephant, peasant workers, the Great Wall of China, the Sydney Opera House and various street scenes. Both of the Voyager spacecraft have now left the solar system, travelling at a rate of 300 million miles a year. It will take them about 40,000 more years before they reach another star. So far there has been no response from any potential recipient.

Another proposal for an interplanetary message has been made by NASA and European Space Agency's spaceflight, Cassini, launched at Saturn's moon, Titan, on 15 October 1997 after a two-day delay due to computer malfunction. This message, which was jointly proposed by the radio astronomer, Gregory Benford and space artist, Jon Lomberg, will be aboard the Titan lander, called Huygens, and was made of diamond which can survive for a billion years to be read if intelligent life evolves out of Titan's primeval soup, or by other interplanetary travellers. In the event, 600,000 signatures were collected and recorded on a CD. One optimistic hope is that it will be read by our descendants when they explore the outer parts of the solar system, but it is more likely that the meaning of such messages is bound up with a desire to ignite public interest in the space programme.

13

It is, of course, extremely unlikely that the messages aboard the Pioneer, Voyager and the Cassini spacecrafts would initiate a two-way dialogue. This was not the intention. Moreover, if any of these messages is discovered and translated by intelligent extraterrestrials in about 40,000 years' time it will hardly be representative of life on Earth as it will exist then. The great mass of the universe is a past universe. We can only observe it as it was. The actual universe is unobservable. Contact with extraterrestrials is unlikely to involve any sustainable dialogue. But one-way contact from the past would still have tremendous significance. Our lives today are very much shaped by messages and statements from the past, including the Judaeo-Christian Bible and other ancient texts.

Developments this century have expanded the scope of the observable universe. Whereas earlier thinkers pondered over the theoretical possibility of extraterrestrial life, SETI investigators now have the facilities to conduct an empirical search. Already SETI research is underway for the detection of micro-organisms in the solar system, with particular emphasis on Mars. By definition SETI is an experimental science. SETI researchers insist that claims about the existence of ETI cannot be substantiated by appeal to theory, no matter how compelling the arguments. Observational developments in twentieth-century astronomy have been enhanced by techniques involving radio astronomy and X-ray astronomy. Prospects of an ETI encounter have been primarily linked with developments in radio communication.

Radio communication

Radio astronomy

Radio astronomy began in 1931 when the US communications engineer, Karl Jansky, constructed a large aerial to track down the source of disturbances to long-range radio transmissions. He discovered radio signals emanating from the centre of the galaxy. The Second World War saw the development of radar, and then the building of the giant radio telescopes. After the Second World War techniques developed which led to the growth of radio astronomy, leading eventually to the discovery of previously unsuspected cosmic objects such as pulsars and quasars, leading to a transformation of our views about the universe, which turned out to be far stranger than had been imagined. The origins of SETI lie in developments in radio astronomy rather than in the space pro-grammes of the 1960s and 1970s. This 'new' field of astronomy recognized that microwaves radiated from objects in space beyond those contemplated in the missions of space probes and far beyond the optical limits of conventional astronomy. In principle, intelligent signals could be transmitted from any part of the galaxy at the speed of light.

X-ray astronomy

This branch of astronomy began in 1949 when it was discovered that the Sun emits X-rays. X-ray and Gamma-ray techniques facilitated observations of neutron stars and even the vicinity of black holes. In December 1970, NASA's Explorer 42, Uhuru, was launched off the coast of Kenya. This small satellite had on board two X-ray detectors which for the next three years scanned the entire sky as it orbited the Earth, revealing countless X-ray sources, some associated with our galaxy, and others with different galaxies. One advantage of the Uhuru satellite was that individual objects could be observed over extended intervals of time. This enabled accurate identification of the optical source responsible for the X-rays. Furthermore, it enabled identification of star masses, sizes and densities, providing evidence regarding their past and future histories. In 1978 NASA's Einstein satellite was launched, carrying an X-ray telescope, thus initiating the era of X-ray astronomy, providing fresh information about quasars, neutron stars and black holes.

Other forms of radio communication

For many SETI scientists sending physical objects to the stars is regarded as too costly in time and energy. For them the best option for contact is by means of electromagnetic communication. The most popular current option for communication with an ETI is by means of low frequency radio waves, which travel at the speed of light and are easy to transmit and receive. The idea of radio communication with ETI gripped two of the early pioneers of radio, Nikola Tesla and Guglielmo Marconi, who listened earnestly for either Venusian or Martian signals. In 1901 Marconi transmitted the first transatlantic radio signal. That same year Tesla recorded rhythmic signals on his radio receiver in Colorado Springs and was convinced that they came from an ETI on either Venus or Mars. In 1902 Lord Kelvin, the British physicist, supported Tesla and announced his belief that as New York was so well lit, it was the best place for Martians to observe and the most likely place for them to make contact. During the 1920s Tesla's rival, Marconi, believed that he too had discovered Martian signals, but in both cases there was no verification. In the early days radio reports of an unknown radio sound were often explained as ETIs. Marconi tended to explain any unexplained radio signals as evidence of ETI. The reasoning is as fallacious as the belief that unexplained bumps in the night prove the existence of ghosts.

There was popular speculation that an ET signal had been detected in 1968 when Jocelyn Bell, who was then a Cambridge graduate student, discovered rhythmic sounds from interstellar space. ETI was ruled out when it was discovered that they were from pulsars, spinning neutron stars from supernova remnants. In fact the first pulsar was named LGM1 (Little Green Man 1). In 1969, when astronomers found a pulsar at the heart of the Crab Nebulae, which was part of the remains of the supernova of 1054, another wave of popular

ETI speculation ended. Many other sounds attributed to ETI have turned out to have come from terrestrial origins: the effects of electrical motors, radio stations or natural signals called 'whistlers' which are produced by distant lightning.

Enthusiasm over alleged ET signals in the early decades of the twentieth century was nourished by Lowell's popular theories about Martian life. It provided the background to the hysterical response to Orson Welles' notorious *War of the Worlds* broadcast in 1938. Even when it was broadcast again fifty years later there were worldwide reports of panic and pandemonium, thus revealing the extent to which popular beliefs are removed from scientific theory. This event was so embarrassing that discussion of ETI fell into disrepute among astronomers. Interest in SETI was not to recover until 1957, when the space age opened with the launching of Sputnik.

The early pioneers of SETI in the late 1950s were Giuseppi Cocconi, Philip Morrison, Frank Drake and Carl Sagan, who marshalled arguments based on what we know about terrestrial life and its origins, and applied them to astronomical data on conditions elsewhere in the universe. In 1959 SETI achieved scientific respectability when the scientific journal, *Nature*, published a paper by the two scientists, Cocconi, a cosmic ray specialist and Morrison, a physicist. This paper addressed the question of radio communication from ET sources. The problem of communicating with ETs is that first there has to be an agreement on a medium, which in the 1950s was radio, and then agreement on a frequency. If it is assumed that ETs want to communicate by radio, then the problem of which noise-free frequency out of millions has to be considered. Cocconi and Morrison proposed a frequency which was based on the rate at which hydrogen atoms emit radiation when the spin axis of the electron orbiting the nucleus flips over from being parallel to the nucleus's spin to being opposite to it. They argued that the hydrogen frequency 'has a universal uniqueness, not set by anthropocentric considerations, that fits it as the outstanding choice for potential communicators who have not had the opportunity to reach agreement on a frequency' (Cocconi and Morrison, 1959: 844). Hydrogen is one of the most abundant elements in the universe; about 999 out of every 1,000 atoms that exist are of this substance. The characteristic frequency for hydrogen is 1,420, 405,752 times per second, which can be expressed as 1,420 megahertz (MGz) – a band in the noise-free area. Another band in the noise-free area represents the hydroxyl radial (OH) and is known as the 'Waterhole'. The hydroxyl radial emits radiation when it changes its rate of spin. Because hydrogen and hydroxyl combine to make water, the frequency corresponding to its radiation emission is known as the 'Waterhole'. The Waterhole has several lines near 1.65 GHz. The letters GHz mean gigahertz, and 1GHz is equal to 1,000 million vibrations per second. In comparison an AM music station is broadcast around 100KHz, which is 100,000 vibrations per second. Because these frequencies have associations with life, it is assumed that intelligent ETs would choose to communicate through them.

COMMUNICATION WITH ETI

Throughout the 1960s Cocconi and Morrison, at Cornell University, employed radio telescopes tuned to the hydrogen frequency to search for an intelligent signal. A similar search was conducted by Frank Drake, head of Project Ozma, so named after the Wizard of Oz. Drake's programme, which began in 1960, used a 26-metre Tatel radio telescope in the National Radio Astronomy Observatory (NRAO) in West Virginia. Project Ozma searched for radio signals near the hydrogen frequency from two nearby Sun-like stars, Epsilon Eridani (10.2 light years away) and Tau Ceti (11.9 light years away). There were no results after 400 hours listening. During the next thirty years over sixty searches, accumulating over 200,000 search hours, in the USA, the former Soviet Union, Australia, Canada, France, Holland, England, India, Japan and the Argentine, were conducted at various 'magic' frequencies. Of these about 90 per cent were conducted on the hydrogen line. One negative search during the 1970s of ten stars on the narrow band 21-cm wavelength was reported by Gerrit L. Verschuur (1980) of the NRAO, who suggested that advanced extraterrestrials might have instituted a series of protected bands for radio astronomy such as the 21-cm wavelength.

While several searches have been funded by governments, others have been financed by private means. Some, the 'dedicated' searchers, have diverted radio telescopes to pure SETI work, while other 'parasitic' searches have examined data supplied by conventional astronomic research; some have hooked up to disused early warning systems for detecting Soviet missile attacks, while others have relied upon portable receivers which they connect to radio telescopes whenever they can beg a few hours. None have reported contact. Although many searches have encountered many unexplained signals, none have been repeated or independently verified. Usually these unrepeated signals are dismissed as glitches in the receiving equipment. However, irregularities in the ionized gas between various stars can either depress or boost a narrow-band radio signal. Thus a signal that is depressed, and only momentarily boosted, would scarcely count as a repeatable event. If this were the case, then it is possible that many signals are being depressed and some of these one-off signals could be part of a continuous beacon. The only solution is to perform many thousands of repeated searches after a reported signal.

The state-funded Project CETI headed by Josef Shklovskii in the former Soviet Union made no progress after listening to 600 nearby stars, and the investigators eventually gave up. Political hostility in the USA prevented direct funding from NASA from the early days of SETI until 1982. Later, SETI's status within NASA improved and the SETI Project Office became a fully fledged department. The longest full-time search to date has been undertaken by Bob Dixon of Ohio State University, who began, in the early 1970s, listening at the 1.42 GHz (hydrogen) frequency, also known as the '21-centimetre line'. Of significance in SETI's history is the famous 'Wow' signal of August 1977 which was recoding by a technician at Ohio in response to an apparent signal. Unfortunately the signal was not repeated. Before the Ohio telescope was

17

terminated in 1998 Dixon had extended his search to cover the Waterhole, up to 1.7 GHz. Since 1983 Paul Horowitz, at Harvard, has been searching various 'magic' frequencies, but has split his frequencies into very narrow bands, about 0.05 hertz in width. The cost of his research was partly met by a grant from the film producer, Steven Spielberg. After five years of continuous sky survey this search found a handful of candidate radio signals with a narrow bandwidth which are not attributable to natural sources or interference. But they have not been repeated. While non-repeatable data is scientifically worthless it does indicate a need for a more comprehensive search. Horowitz is currently leading Project META (Megachannel Extraterrestrial Assay) at Harvard, which is supported by the Planetary Society.

Two assumptions can be discerned in these radio searches. First, the transmitting civilization will be more advanced than us; if they were not, then they would not have the appropriate technology. If they do have the technology, they are very likely to be more advanced as our communicative history is so short. Radio communication on Earth is less than 100 years old and has almost reached technical perfection – a state in which further technical refinements will not improve results. If radio communication can be as good as it can get within such a short time it is likely that other civilizations have either got it perfect or not at all. Second, the transmitting civilization, being more advanced than ours, would do their best to make it easy for us to listen. This is why they are likely to choose one of the various 'magic' frequencies in the noise-free area. This assumption provides a powerful heuristic limitation on the scope of the search.

Most proposals for radio searches are committed to receiving information rather than broadcasting messages. This is bound up with the belief that the transmitting civilization will be technologically superior. They will discover us, just as Columbus discovered America; the native Americans did not discover Europe. So instead of sending them signals we should await theirs. It is also more economical with resources to receive signals rather than broadcast. But if they are so much more advanced, what is their point in communicating? And could they communicate even if they are intelligent? They may be like the ancient Greeks; intelligent but lacking the technological means to communicate over long distances. Extraterrestrials outside of our technology threshold may be intelligent but we can have no way of confirming it if they have no access to a radio transmitter.

It should be stressed that persistence, despite repeated failure, is not a sign of dogmatic resistance to falsification. In this respect SETI could be described in Feyerabend's terms as a research programme requiring breathing space. The technology is in its infancy and only a micro-fraction of the space has been searched. Nevertheless, SETI researchers do resemble the old panhandlers who persistently claimed that the valuable nuggets were to be found in the next search. And, like the panhandlers, they are aware that it will require only one verified find for them to be showered with reward. Drake (Drake and Sobel, 1993: 206) may have overstated resistance to falsification when he said: 'No amount of failure in SETI endeavours constitutes proof that we are alone. Only the existence of life is demonstrable.' Despite the fact that SETI resembles a search for a very small needle in a very large haystack there must, nevertheless, be criteria which set a limit on failure. At present, however, the search has barely begun. Even a massive search, lasting hundreds of years, could miss technological civilizations which emerge with a capability for radio communications during the search period. It is difficult to draw the conclusion from a negative search that the target areas are devoid of life. One can only conclude that during the time of the search no signals were being directed at us. It is worth noting that the Earth has only been radio-detectable for a century. Thus a radio search of Earth before the late nineteenth century would not reveal 'intelligent life' whereas a search during this century would reveal to any extraterrestrial civilization that there is intelligent life here.

The fact that we are now detectable may stimulate our galactic neighbours into communicative efforts. The fact that we are of potential interest should stimulate us to further efforts to listen. Suppose that, like us, ET civilizations are listening and have powerful receivers which can pick up our leakage signals from radio and TV broadcasts, bearing in mind that radio and TV signals are weak and are not broadcast on any of the magic frequencies. If the nearest receiving civilization is 40 light years away they will have picked up our leakage signals beginning in the 1930s which have continued since then and we could expect their reply from the year 2010 onwards. On this view the passage of time increases the prospect of contact. But it also means that there is a reduced likelihood of a receiving communicating civilization nearer than 40 light years away and that this distance will increase with the passage of time. Meanwhile, better technology and bigger searches using some of the world's largest radio telescopes are currently underway.

Why should they wish to communicate?

Most SETI researchers assume that ETs are willing to communicate and share their superior knowledge with us. But why should they wish to communicate? What likely benefit would accrue to them? One suggestion is that they realize that their sun is nearing its end and with it their whole civilization. They transmit, giving details of their history, thus expressing a desire that some memory of their existence will survive them, with no hope of receiving a reply. A second suggestion is that a civilization begins space exploration and discovers reasonable proof that life has evolved on another nearby planet, thus giving them optimistic reasons to initiate a communicative network with other civilizations. A third suggestion is that two advanced civilizations may have evolved close to each other so that first efforts are immediately successful. This would generate an intense search from the beginning. Once communication is established, it would then be likely to be widespread, as success would encourage more beacons. Thus, so the arguments go, if interstellar communication exists, it is probably a reality for countless races who are participating in a galactic network. The assumption that ETs have a desire to communicate has meant that searches have been designed to identify a deliberate signal. This may limit the prospect of detecting intelligent life as there may be technological civilizations who have no conscious desire to attract our attention. A case can therefore be made for eavesdropping on their radio and TV leakage. Thus Woodruff T. Sullivan (1980) argues that a well-designed search system must allow for radio leakage as well as an intended message. This may be more difficult to detect, but leakage can be separated from natural emissions, so long as it is narrow band and periodic.

What kind of intelligence is sought?

The technology employed to conduct the search for ETI has imposed its own definition of intelligence. SETI researchers are looking for a technological intelligence appropriate to the late twentieth-century scientific civilization on Earth. Given the potential for a vast diversity of intelligent cultures, this might appear as a rather narrow and restricted objective. But this is all that existing SETI technology can detect. There is an anecdote told by SETI scientists: a man looks for a key that he has lost in a dark street. There are a few lights, so he concentrates his search beneath the light because that is all he can do. The light provides the reasonable range of the search. Likewise SETI technology provides the reasonable range of the search, and confines it to the search for similar technology. There is, however, a further reason to suppose that ETI will share an understanding with terrestrial technological intelligence. Given that curiosity is a related feature of intelligence and the same physical laws operate throughout the universe, it is likely that curious ETIs will have discovered what we have discovered.

Many of the assumptions of our scientific civilization are accordingly built into the SETI programme, such as the assumption that nature itself is not intelligent. Data that have a natural origin are excluded. SETI research is primarily a search for evidence of technology. Some SETI scientists have sought evidence of interstellar nuclear waste dumping, looking for evidence of leakage from ET nuclear plants by means of the detection of tritium in the vicinity of a solar-type star. This radioactive isotope is a particularly good indicator of nuclear technology; it has a half-life of 12.5 years and its presence would indicate leakage from a nuclear fusion plant. Presumably any evidence of an ET Chernobyl-type disaster would be an indication of intelligent life! If detected in our solar system tritium could indicate the presence of a nearby ET probe.

Current radio astronomical searches are based on a concept of high technology intelligence and are unlikely to make contact with any life-form that has not evolved such an intelligence. They will not make contact with any equivalent of ancient Greece or ancient China. SETI scientists define intelligence in terms of the power of abstract thought, the ability to construct at least partly successful models of the external world and the ability to use these skills to build things and, in some cases, predict the future (Gribbin, 1991). A related definition of intelligence has served as a guideline for the Harvard-based SETI project: 'life that has acquired a technology required to carry out interstellar communication' (Horowitz and Alschuter, 1990: 261).

What kind of signal is likely to be transmitted by intelligent ETs operating with similar technology to our own? One of NASA's radio astronomy experts in SETI, Jill Tarter, has specified the requirement that:

> the candidate signal be narrowband or concentrated in frequency, and perhaps also concentrated in time. If the latter is the case, then the signal must be no more extended in frequency than is demanded by its time duration and the laws of physics as we now understand them.
>
> (1991: 191)

What frequency, or range of frequencies, should be searched? A meaningful signal cannot be identified against a noisy background. This eliminates long radio wavelengths. At long wavelengths (approximately one metre) noise background is too great, owing to a process known as synchrotron radiation. On the other hand, very short (higher) frequencies (near one millimetre) exhibit too much noise from the remains of the big bang that created the universe. In the infrared region, at wavelengths of the order of one micron, there is a strong background noise from the glow of warm dust clouds. At the shortest wavelengths of all, X-ray and Gamma-ray emissions create background noise.

The quietest space is in frequency units known as the microwave portion of the electromagnetic spectrum. In frequency units this extends from about 1GHz to 10GHz which includes the hydrogen frequency and the Waterhole. For SETI researchers the range of frequencies relatively free from cosmic noise is known as the Free Space Microwave Window, representing a relatively clear channel for any species, anywhere in the galaxy, that can operate a radio transmitter.

Project Cyclops

In 1971 a NASA team headed by Bernard M. Oliver, Vice President of the Hewlett Packard Computer Company, proposed a sophisticated system for listening for an ETI radio signal. It was published under the title *Project Cyclops: A Design Study of a System for Detecting Extraterrestrial Intelligent Life* (Oliver and Billingham, 1973). The Cyclops proposal would have involved a system of 1,000 antennae connected by information tunnels to a central information-gathering building. It was to be highly automated, requiring human attention only if an anomaly was found. Staff and their families would live in Cyclopolis, a town to be constructed with schools, housing and stores. Although Cyclops was designed as a receiving system, there was nothing in its design to prevent it from being used as a beacon. It was considered a good idea to alternate between listening and transmitting; if the proposed target search of 1,000 stars produced no signal, then a brief message could be broadcast for a year or so before moving the search into deeper space.

Project Cyclops recommended the microwave window as the most suitable part of the spectrum for the search and outlined the design of a data-processing system capable of a simultaneous search of the most likely area within this window. A blind search was ruled out and Cyclops envisaged initial target lists of stars which would expand and be subject to re-evaluation as construction of the complete system developed towards a total fully automated search. Some places, however, were deemed to be better to search than others: for example, if interstellar communication is a reality, then there may be a special place in the galaxy where the beacon is likely to be located. Thus, held the Cyclops team, it is likely that the galactic culture has constructed a powerful beacon in the centre of the galaxy, or maybe members of the galactic culture have adopted a convention of radiating beams in a specific direction, either from or towards the centre.

The Cyclops team expressed a preference for a radio search at the lower end of the microwave window, between 1-3 GHz. Their reasons were that there are smaller Doppler shifts; less stringent frequency stability requirements; a greater collecting area for the narrowest usable beam; reduced cost per unit area of collecting surfaces; smaller power densities in transmitter tubes, waveguides, feeds and/or radiators, thus allowing higher powers per unit, greater freedom from O_2 and H_2O absorption, which may be more on some planets having our life-forms.

What kind of radio signal did Cyclops envisage? The underlying assumption was that the extraterrestrials would go to great lengths to make the message as decipherable and as obvious as possible. The beacons would have the following properties: there would be continuous transmission; they would be aware that those listening may ration time for each star, so that they would signal continuously to avoid being missed; they would be monochromatic; they would probably be circularly posed; their radio waves would arrive with either right- or left-hand polarization, thus reducing the need for extra equipment by the searching race. Finally, they would be information-bearing, not just a signal awaiting a reply, which would waste too much time; they would start with information, including information on how best to reply, explaining which frequency or modulation rate was preferable. The authors of Project Cyclops also recognized the problem of communicating with beings who may not share similarities in natural history or culture; they recognized, from terrestrial examples, that there is a great difference among races as to what is considered obvious. Mathematics, they held, has a universality which would have a meaning for any society capable of radio communication, and the Cyclops team suggested that ETIs might attempt a pictorial message in the form of a binary code of sequences having particular lengths that could be arranged into a picture. There would, however, be a major problem of interpreting the picture without a common set of rules or conventions for recognizing and reading pictures.

Project Cyclops was not deployed and, with an estimated cost of $6 billion (roughly equivalent of what the USA paid for three months of the Vietnam War), was rejected. But the report indicated that the technology does exist for an effective search for ETI, and that were such a system tried for a given period without success its negative findings would count as a reasonable falsification of pluralism. Some of the leading ideas of Project Cyclops were to be employed in NASA's High Resolution Microwave Survey (HRMS), which was formerly referred to as the Microwave Observing Project (MOP).

NASA's High Resolution Microwave Survey and Project Phoenix

On Columbus Day in 1992 NASA scientists began a search for ET radio signals. The search was originally scheduled to run for ten years with an upper budget limit of $100 million dollars. It was described as part of the US commemoration of the discovery of America, looking towards its future as well as its past. The search was to be conducted primarily by means of radio telescopes and involved two complementary teams. The first team, based at the Ames Research Centre, were to employ a 'Targeted Search' to examine at high sensitivity about 800 stars like our Sun which are located within 80 light years from Earth. It began with a concentrated search targeted on a star in the Constellation of Opiuchus. Using the largest antennae available, including the 305-metre dish at Arecibo Observatory in Puerto Rico, each star was to be carefully scrutinized on frequencies between 1GHz to 3GHz in the microwave band, which would include the Waterhole. Radio telescopes at various sites throughout the world were to be programmed to follow each designated star for several minutes at each frequency while the 'targeted search analyzer' was to search for complex signal patterns. The assumption underpinning the Targeted Search was that intelligent civilizations are so common throughout the universe that our nearest neighbours may be within 80 light years of the Sun.

The second search was based at the Jet Propulsion Laboratories in Pasadena, California, employing a 34-metre radio telescope at the Goldstone complex of telescopes in the Mojave Desert, California. This was known as the 'All Sky Survey'. It was to be less sensitive than the Targeted Search but the sacrifice of sensitivity allowed greater sky coverage. It assumed a smaller number of communicating intelligences, although Frank Drake saw this as the most likely to succeed. While sacrificing sensitivity, the All Sky Survey was intended to employ a search covering 99 per cent of the heavens ignored by the Targeted Search. This was a much wider search, using three 34-metre dishes of the world-wide NASA Deep Space Network; it was intended to tune into frequencies that cover almost the entire ground-based (1–10GHz) microwave window: hence the original title Microwave Observing Project (MOP).

Thus conceived, the dual project had an advantage over many of its predecessors because recent computer technology has made it possible to listen to many frequency channels simultaneously. Data from 20 million frequency channels were to flash through digital processors and several dozen of the world's most powerful super-computers (including the CRAY X-MP/18) were required to match the speed of this hardware, completing tens of billions of mathematical computations each second. Multi-Channel Spectrum Analysers (MCSA) will process 25 billion bytes of data each second, and were to be programmed with signal-recognition algorithms to differentiate potential ET signals from the confusing background of terrestrial and cosmic noise, recording only the most promising data for follow-up analysis. Early tests gave grounds for optimism regarding the system's ability to isolate signals from the background noise. On trial runs, signals from the Ulysis spacecraft were picked up as it approached Jupiter. It was to be the most comprehensive search ever. In the first minute of its operation it had searched more than all previous searches combined. The scale of the search can be appreciated when we consider that every second the NASA instruments could process information equivalent to several *Encyclopedia Britannicas*. The SETI Continuous Wave Detector can make 1 billion tests per second, which is equal to scanning the entire *Encyclopedia Britannica* each second to pick out a significant three-word phrase.

Among the spin-offs from this research could be the data-reducing hardware and software which could initiate a revolution in computer design. So far NASA searchers have devised silicon chips that can analyse millions of frequencies simultaneously, and software that can search for different kinds of signals, including continuous tones, pulses, or combinations of both. Algorithms, developed for the detection of ET transmissions, may have other applications including tests for the sensitivity of computer searches in the diagnosis of breast cancer.

Unfortunately the HRMS suffered a serious setback in the 1993 US budget cuts when Congress killed off the All Sky Survey when it was less than a year old and had only covered one-thousandth of the intended search. The 'flying saucer' image of SETI, however misguided, encouraged politicians to withdraw financial support for the HRMS. It is fashionable among 'serious' people – who include politicians and members of the media who are close to them – to portray beliefs in the existence of ETI as a manifestation of gullibility, superstition and lack of education. However, the Targeted Search managed to escape the cuts and survived, as the SETI Institute, which is responsible for it, is independent of the government and was able to solicit outside funds. It consequently survived, and $4.4 million was initially raised for the Targeted Search, with substantial backing from four US high-technology entrepreneurs, including David Packard and William Hewlett, founders of Hewlett Packard, Gordon Moore, chairman of Intel and Paul Allen of Microsoft. Arthur C. Clarke also made a 'significant' donation (Anon, 1994). Its current name is 'Project Phoenix'. Throughout 1995 and 1996 Phoenix continued, committed to a search of 1,000 nearby Sun-like stars. The SETI Institute was, however, able to use the $58 million of government investment in equipment. Even in its reduced size it is still the most comprehensive search ever undertaken. Early work in 1995 involved a five-month observation of the southern sky with the Parkes 210-foot diameter radio telescope in Australia, which focused on 200 Sun-like stars. After the Australian search the Phoenix receiving equipment was stationed at the 1,000-foot diameter Arecibo telescope in Puerto Rico until the end of the decade. The frequency coverage of Phoenix is the quietest part of the microwave window, between 1 and 3 GHz.

NASA's SETI collapse did not herald the demise of the search: on the contrary, SETI projects are springing up around the world. Harvard University and Ohio State University, which have well-established independent programmes, were joined by others in the Argentine, France, Australia, Italy and India.

One of the current searches is the University of California's SERENDIP III, which has been 'piggy-backing' on the Arecibo telescope in Puerto Rico. The University of California search is using the SERENDIP III spectrum analyser, which examines 4.2 million channels every 1.7 seconds in a 12 MHz-wide band. The Argentinian Institute of Radio Astronomy has two 30-m radio telescopes built to study southern hemisphere galaxies. It has operated a listening programme since 1986. Among the European searches is one led by François Biraud, who has been using the radio telescope at Nancy since 1981, and is monitoring 300 of the closest Sun-like stars. In Italy the Large Northern Cross, at Bologna, is sweeping the sky's northern hemisphere.

With off-the-shelf equipment the modern SETI amateur can search as much as the professionals did twenty years ago. SETI is rapidly becoming a branch of science that can attract low-budget researchers. Moreover, if amateurs link up they can obtain a much larger field of view. Hence the SETI League, which was founded in 1993 after the withdrawal of US government funds from NASA's search, originally intended to have over 5,000 amateur searches underway by 2001. This was known as Project Argus, after the giant from Greek mythology who had 100 eyes. On 16 May 1998, a project called SETI at Home was initiated, with an original objective to involve over 100,000 volunteers who would scan data from radio searches with their PCs. SETI at Home captured the public's imagination, with numerous individuals and schools participating. By the beginning of 2000 over 1 million people were involved with the project, and by May of that year it had attracted over 2 million participants.

Yet these searches can only cover a fraction of the available space. For reasons of budgetary control the original HRMS was to be conducted below 10GHz (one-tenth of the microwave window) and limited to operations on the surface of the Earth. Because of this restraint on location, a search above the 10GHz limit would be frustrated by considerable background noise from water vapour and molecular oxygen in the Earth's atmosphere. Until the search can be conducted from stations located in space the best available is the ground-based portion of the microwave window. But even within these parameters a systematic search of every possible frequency in the 1–10GHz range is impossible. Suppose the channels 1–10 are broken up so that each could be listened to simultaneously.

The problem is that within the 1–10 window are 9,000 million possible 1Hz-wide channels, 4,500 2Hz-wide channels and 9 million 1KHz-wide channels. With present technology it is impossible to manufacture a spectrometer with 9,000 million channels, although the current goal of SETI research is to produce one with a capacity for 10 million channels. But even this number cannot be adequately searched simultaneously. So current technology is geared to better pattern recognizing software to reduce the scope of the search and computers designed with greater speed and larger memories. Thus efforts towards greater computer speed together with facilities to 'recognize' interesting patterns which may be indicative of ETI, will be an important feature of the project. A new generation of super-computers with parallel search and enhanced memory might just take SETI to the starting point.

If an intelligent signal is received, scientific methodology would require its verification and repeated confirmation of it. Previous searches have recorded dozens of unexplained signals but none have been repeatedly observed. The signal must also be independently detectable by other scientists. This is partly because of the need to protect against hoaxes. Once confirmed, the SETI protocol requires that, first, another researcher with suitable equipment that can confirm its authenticity is contacted. Second, the International Astronomical Union and the Secretary General of the United Nations must be informed. Third, a press conference is to be called, ensuring that the information is disseminated correctly. Fourth, if the signal can be translated, the protocol warns against an immediate reply: 'no response to a signal should be sent until appropriate international consultations have taken place' (Henbest, 1992a: 13). The International Academy of Astronautics in the USA has produced a document entitled the *Declaration of Principles Concerning Activities Following the Detection of Extraterrestrial Intelligence*, which requires global cooperation in the formulation of a response to a signal or landing.

It is expected that the report of a contact would occupy newspaper headlines for a few days and then be quietly forgotten. Apart from a few sensational newspaper stories – of which there are many already – the full consequences of contact might not be very dramatic in the early stages. There would be excitement among the scientific community in response to the recognition that it was an artificial signal. This would be followed by the major difficulties encountered when trying to interpret it. (We might not even know what they mean by intelligent communication.) During this period we would be shielded from the full culture shock anticipated with the discovery of ETI. Like many great discoveries, knowledge of the existence of ETI may be either initially ignored or misrepresented, but gradually accepted as other beliefs are transformed. Just as it took time to fully accept the existence of virus or black holes, acceptance of ETI is likely to accumulate as bits of evidence emerge together with a gradual shift in beliefs. The long-term effects, however, are considered to be profound. After the first authentic signal many others would follow because once there is success the search will intensify. Optimists would look forward to dialogue with an advanced ETI who could solve many of our problems and invite us to join the Galactic Club, while pessimists warn of invasion and slavery. But what if no signal is encountered? Demands will inevitably be made for larger and more extensive searches with radio telescopes located on the Moon at a cost exceeding $100 billion, for there can be no conclusive proof that the search has failed. Either there is a pot of gold awaiting the SETI teams or a journey down an endless road.

Despite the advanced technology, radio searches face tremendous problems. The terrestrial use of microwaves is on the increase and the consequent background noise has become a matter for serious concern. Radio frequency interference (RFI) is increasing from the use of cellular phones, microwave data transmissions, police radar and communications satellites. Over 100 new satellites are launched each year, many of which leak signals into the spectrum reserved for radio astronomy. One serious example of radio interference was the former Soviet GLONASS navigation satellite system which, for a period after 1982, was a considerable source of interference for radio astronomy, especially for receivers on the hydroxyl line. During the 1980s transmissions from this system affected 1612 MHz OH observations all over the world. This particular problem was resolved after talks between the GLONASS administration and the Inter-Union Commission on Frequency Allocation for Radio Astronomy and Space Science (see McNally, 1994). In July 1999, astronomers presented a document seeking protection for radio astronomy to the UN Committee for the Peaceful Uses of Outer Space. It requested that light sources, radio transmissions and various space junk, should be classified as global pollutants, and that any planned space activity likely to affect astronomy should undergo prior evaluation. Unless urgent curbs are introduced to protect against RFI and other forms of space pollution, it may well be the case that our technological civilization has paradoxically foreclosed on radio communication within a little over the one hundred years it has been available.

SETI researchers are not merely looking for a needle in a haystack; they are looking for a very special needle in a haystack which is already full of needles. Or, to use another image: it is like trying to hear a pin drop in a crowded ballroom with more dancers arriving all the time. Some pessimistic forecasts suggest that the present rate of increase of human-made radio noise will put ground-based radio astronomy out of business within twenty years. As powerful interests are involved there is no strong political will to curb RFI. The future of radio astronomy in general, and SETI in particular, lies in operations conducted in outer space or on the far side of the Moon.

Optical signals: an alternative

Questions have been raised with regard to SETI's reliance upon radio signals as a means of communicating with ETs. As early as 1961 it was proposed that optical signals may be an alternative to radio waves. The search for evidence of extraterrestrial communication by optical masers was first proposed by R.N. Schwartz and C.H. Townes (1961) in the journal, *Nature*. They suggested that an advanced civilization might have become very sophisticated in the use of optical masers rather than in the techniques of short radio waves. Nigel Henbest (1992b) revived the debate over optical signals when he argued that radio communication in space is very much a 1950s' creation, when the idea of interstellar signals was associated with powerful radio telescopes. But communicative technology has moved on during the past forty years and a more advanced civilization would have moved on even further. The future for interstellar communication, argued Henbest, lies in lasers. It is likely that a more advanced society will be thinking of laser rather than radio.

In 1974 there was a small-scale search of three stars which used the Copernicus satellite to look for ultraviolet laser emissions of intelligent origin. Nothing was found. Nevertheless, the case for an optical search rests on an appeal to the technical development in lasers during the past forty years together with the possibility of setting up telescopes beyond the Earth's atmosphere. Lasers are amplifiers of radiation which send out extremely 'clean' light signals of a very small bandwidth. Their light is not diffused or fanned out like a searchlight. J.D.G. Rather (1988: 386) has provided four arguments in support of an optical search. First, lasers are 'a versatile technology for producing highly directional beams that deliver maximum power on target'. Second, 'by targeting the power, rather than broadcasting it, vastly better use is made of the energy available'. Third, 'the availability of laser power at any chosen wavelength makes possible the choice of wavelengths that show the intelligent signal in maximum contrast to its natural environment'. Fourth,

> the potentially enormous brightness (watts/steradian) of a well-designed laser system makes possible delivery of high signal-to-noise signals having tremendous data rates anywhere in the Galaxy. There is no need to worry about photon noise when dealing with truly advanced civilizations.

The instruments are available for an optical search, but until fairly recently it has been assumed that those who transmit would opt for microwave transmitters which require less power than optical wavelengths. A search for optical lasers was considered during the 1970s and 1980s, but was rejected by many SETI scientists, who favoured microwave searches on the grounds that microwaves offered a maximum detectable range because they contain more photons per unit of energy. Laser searches also faced problems with the construction of receivers, which were said to require highly polished antenna surfaces that are susceptible to weather changes. Interstellar dust may also interfere with optical signals. The Cyclops study made a careful comparison of lasers versus microwaves and came down in favour of the latter. Primarily, the reasons cited were that lasers would consume more energy and were prone to interference from synchrotron radiation, thermal noise, quantum noise and star noise. The authors of Cyclops put forward the case for a search in the free space microwave window between 1–60 GHz.

Initially, then, the SETI community were unfavourable towards optical searches, but lately the Planetary Society and the SETI Institute have adopted a more positive standpoint. The Columbus Optical SETI Observatory is committed to a laser search, which has been endorsed by Arthur C. Clarke. In 1998 the SETI Institute initiated 'Optical SETI' which searches for pulsed laser signals. This development was facilitated by the promise of a high-intensity pulsed laser which, working together with a modern telescope, could transmit signals beyond 1,000 light years. The argument is that if such a system is possible on Earth, then ETs with similar technological skills might already be transmitting laser signals. The objective of Optical SETI is to employ Harvard's 61-inch optical telescope in a radial velocity of 2,500 nearby solar-type stars. Two teams are associated with this project: Paul Horowitz and his team at Harvard, and one led by Dan Werthimer at the University of California.

Among the reasons for the revival of optical searches is that recent calculations have indicated that ionized hydrogen in the interstellar medium can cause a dispersion that degrades transmitted radio signals, and it may be the case that a more advanced extraterrestrial civilization will be aware of this fact and consequently prefer to communicate via optical signals. Moreover, it has been recognized that laser detection does not require very sophisticated equipment: primarily it requires a pair of fast photon-counting detectors working in 'coincidence', argues Horowitz (2000: 10). One of the problems with optical signals could be the requirement to separate the luminosity of the beam from the background light of the parent star. But even with Earth's current level of laser technology a beam could be sent that 'would appear about 5,000 times brighter than the background light from our Sun', Horowitz points out (ibid.: 10). In which case an extraterrestrial society, at the same level as us or only slightly more advanced in laser technology, would have no problems in producing a beacon that would outshine its parent star.

How would an extraterrestrial society transmit a laser signal? They might either send out a series of laser pulses of considerable intensity or they might send out a continuous beam. Horowitz and the Harvard team employ a system to detect a pulsed laser signal using Harvard's optical telescope. From October 1998 to November 1999 they observed about 2,500 stars without finding evidence of intentional laser signals. But the search has barely begun. With further collaboration these optical searches will extend in both their scope and accuracy. Plans are underway for an 'All Sky Optical Search', using over 500 detectors, which could cover the full northern-hemisphere sky within 150 clear nights (ibid.: 14).

Searches are also underway for a continuous laser signal. Planet finder Geoffrey Marcy is conducting a search at the Lick Observatory in California, the Keck Observatory in Hawaii and the Anglo-Australian Observatory in Australia.

This search is examining the vicinity of 1,000 stars for evidence of intentional narrow laser-like signals.

A step-by-step approach

The main problems with both the microwave search and the proposals for an optical search is that they are high risk methods; they can only be successful if we are listening at the precise time that they are transmitting. Moreover, the longer the search continues without results – despite valid claims that only a small fraction of space has been searched – the risk of public disenchantment increases. The way forward was outlined in the late 1980s when Bruce Campbell (1990), who had been developing techniques for observing planets, suggested that radio searches should be combined with observational searches for planetary systems. He described this as a step-by-step approach. The first step, involved a search for ex-solar planetary systems by extending and improving planetary detection techniques. This meant that larger planets would be discovered first.

The second step depends upon further developments in detection techniques which may lead to information about smaller planets, such as their frequency, whether they have sufficient mass to retain a dense atmosphere, and whether they are located in a thermally habitable zone. The third step would involve a search for conditions favourable to life, by seeking signs of oxygen, whose presence may be an indicator of plant life. A step-by-step approach may not lead to direct contact with intelligent life and may only yield knowledge of primitive plant life. But this would be of enormous significance. This approach is certainly compatible with radio searches and optical searches and, as early steps are taken, with the promising discoveries of large ex-solar planets since 1995, there would be good reasons for devoting longer listening periods in the vicinity of particular stellar systems. Thus further stages in the step-by-step approach could be followed up by greater concentrations of listening zones where potentially habitable planets are predicted.

Neutrino searches

Searches for ETI have to be based on known principles of science and technology. This is one of the ground rules which separate science from fiction, and such rules are the subject matter of philosophical inquiry. According to J.G. Learned, S. Pakvasa, W.A. Simmons and X. Tata (1994: 328), a useful SETI strategy would be to search for neutrino timing signals from an advanced intergalactic civilization. They suggest that 'an advanced civilization colonizing our Galaxy will be likely to have the need to synchronize clocks over interstellar distances'. If this civilization is to maintain communication with its colonies it will 'require interstellar time standards at the limits allowed by physical processes' (ibid.: 321). They propose that this civilization communicates with its colonies with the help of neutrino timing pulses, a system of 'timing marker pulses as narrow as 10^{21} sec' (ibid.: 321). These neutrino timing pulses can be produced with briefer timing and greater luminosity than pulses of electromagnetic radiation. Learned *et al.* point out that the 'production and decays of W and Z° bosons are the fastest processes known in physics' (ibid.: 324) and they suggest that it is likely that an advanced intergalactic civilization will employ timing pulses using electron neutrinos from the decay of Z° bosons. Not only are they of shorter duration and greater luminosity than electromagnetic pulses, they can penetrate and still remain bright; they would not be blocked by interstellar dust or be dispersed in space. Neutrino signals would also be extremely distinctive and are unlikely to be mistaken for naturally occurring processes. Z° bosons could be created by colliding high energy electrons and positrons head-on. Thus giant particle accelerators would be employed as transmitters. But in order to transmit signals across the galaxy these transmitters would have to be the size of the Earth.

Obvious objections can be made regarding our lack of any idea as to how these accelerators could be assembled. But let us assume, for the moment, that advanced extraterrestrials have overcome all technical obstacles. How could we listen in to these neutrino messages? The decay of the Z° boson produces equal numbers of all three types of neutrino, one of which – the electron anti-neutrino – could be detected by a receiving beam of high energy electrons. If this beam contacted the incoming electron anti-neutrinos, then W-bosons would be created. If an advanced civilization is transmitting, they could be detected on Earth by the next generation of neutrino telescopes, some of which are located beneath the oceans. For example, Learned *et al.* suggest that the Deep Underwater Muon and Neutrino Detector (DUMAND) which is being constructed beneath the ocean, could receive a signal from a neutrino transmitter within a distance of 3,000 light years. So if they are busy colonizing, then their timing methods may be detected on Earth.

The technical problems, however, are enormous. There is no plausible description available concerning the construction of the accelerator-transmitter. But if such an object was constructed we might very well be able to observe the transmitter, not merely its messages, from the infrared energy. In fact a civilization capable of producing an intergalactic neutron transmitter would, like Kardaschev–DysonType III civilizations (see Chapter 7), require enormous amounts of energy that could be detected in a search for infrared sources.

Decoding the message

Cocconi and Morrison (1959) suggested that the opening of the message would involve some attention-seeking device, something obviously artificial, like a sequence of pulses representing prime numbers. Then the message might contain familiar information, such as well-known scientific laws. Whatever form it takes, decoding an ET message could be a major undertaking. Until the discovery of the Rosetta Stone, European scholars devoted over a century of wasted energy trying to decode ancient Egyptian hieroglyphics. Some ancient languages, such as the Glyphs of Easter Island and the writings of the Mayas, are still uncoded. Yet these languages reflect a similar biological form and territorial niche in the universe to us. How much harder will it be to translate messages from a distant galaxy from beings of whom we know nothing?

There have been a number of attempts to devise basic systems of communication that can be shared by different beings across the galaxy, but all so far are badly flawed. We only need to consider the difficulties humans encounter when trying to communicate with dogs – a species that has evolved alongside humans on Earth since the hunter–gatherer stage of human existence. Volumes of books have been written on how to communicate with pets who share our homes, yet much of their lifestyle and ways of signalling to each other are beyond our comprehension. If it is so difficult with members of our extended 'families', consider how much more difficult it will be with extraterrestrials. Frank Drake (Drake and Sobel, 1993) recalls a simple black and white message, which he devised, of zeros and ones, consisting of 551 characters, being the product of two prime numbers, 29 and 19. The characters formed a picture which gave information about our solar system, diagrams of oxygen and carbon molecules, the shape of human beings, and representations of humans in binary code. He sent this 'simple' message to other scientists engaged in SETI research and to several Nobel prizewinners. None of them interpreted the message as intended by Drake, although a year later an electrical engineer who was an amateur code-breaker, successfully deciphered it. He was the only one to do so!

In 1960 the mathematician, Hans Freudenthal, devised a language for intergalactic communication which he called Lincos. Lincos is an artificial language, not designed to be spoken, which, in contrast to all natural languages, avoids irregularities and exceptions. Based on pure logic and mathematics, without any cultural accretions, it is built up in the form of a dialogue based on signs; pointing to objects, repeating the use of symbols in various contexts to reduce uncertainty. For example, in the dialogue between A and B, A asks: 'what is $2 + 3$?', and B answers: '$2 + 3 = 5$', and A says 'correct'. Then A asks 'what is 10×10?'. B says '20', so A says 'incorrect' and so on. This exchange is designed to convey the meaning of words like 'correct' and 'incorrect'. The mode of communication was intended to be via radio bleeps, which would indicate mathematical symbols, rather than by ostensive behaviour. Given that we have little idea of their physical structure, environment and culture, more sophisticated forms of communication could be withheld.

In a review of Lincos, S.W.P. Stern (1962) suggested that communication should be limited to mathematics until we are sure what kind of beings we are dealing with. But how do we know that other intelligent beings will appreciate mathematics? If they have memories and can perceive the present falling into the past, says Stern, they will have the minimum requirement for developing mathematics. But even their mathematics could be different, and it is not obvious that any form of applied mathematics would have any similarity. It is unlikely that applications of maths to commerce, banking and gambling would be found among ETIs. The major problem with Lincos and other basic languages is that no message, however simple, can carry its own interpretation. This would seem to be a problem facing all cosmic-style languages.

Lincos is based on the assumption that mathematics is foundational in the acquisition of language, and the designers of Lincos and its successors maintain that its foundational concepts require no definition and must be truly universal. All of this rests on the belief that mathematics is truly universal and that language is based upon it, which flies in the face of much of twentieth-century philosophy of language where it is held that linguistic practices are non-foundational. The Lincos system depends upon numerous universals in mathematics, chemistry, physics and biology. But the problem is that diagrams, figures, etc., are representative and carry no guarantee that an alien intelligence will grasp their meaning. Symbols in themselves are meaningless and are not self-explanatory, requiring an interpretative framework and a shared background of tacit knowledge. This is unlikely in the event of first contact with an alien intelligence.

It could, however, be argued that a message transmitted by means of radio waves should in principle, be translatable as there would be a sufficient overlap in meanings displayed in the capacity by both sides to build and operate transmitters and receivers. There is, additionally, a possibility that communicative symbols could be supplemented with pictograms and automatic computer image recognition technology (Breuer, 1982: 134–5). Pictures are an economical medium of communication, and it is possible for certain abstract concepts to be communicated by pictures, providing that those receiving them have optical sense organs.

The controversial 3-minute message that Frank Drake sent out in 1974, from Arecibo to M-13, which is 24,000 light years away, consisted of 1,679 binary pulses arranged in a pictogram, which describes the biochemistry of man and other information regarding our location in the solar system. Providing that the astronomers of M-13 aim their telescopes at Earth for three minutes during the sixteenth of November in the year 25,974, they should receive the message. However, a search among stars which are already within range of the Arecibo radar beams, which has been proposed by Jean Heidmann (1997: 178), could well indicate that replies are being sent. The problem is that by the time the message reaches its destination the signal will have mingled with dust and starlight which will block off part of the message and add noise. An attempt to resolve the problem of an incomplete message was tackled when Lincos was revived during the 1990s by the Canadian scientist, Yvan Dutil, at the Defence Research Establishment near Quebec, and supported by an American organization called Encounter 2001, which sent out a 4-hour-long radio message on 24 May 1999. This message contained around 300,000 bits of information arranged in such a way that knocking out a particular bit should not detract from the overall level of information. The message has now travelled beyond the solar system.

Communicative problems – first contact

What kind of first contact is likely between Earth civilization and ETIs? They might come here unannounced and without warning. This could be indicative of desperation, fleeing from ecological disaster, a dying stellar system, or an act of stealthy aggression. In either case they would come armed and it is very likely that their presence would initiate a violent struggle. We might, at some future date, go to them equally unannounced, and invade them, initiating another struggle. These considerations aside, it would make sense, in terms of resources, for any civilization contemplating physical contact, to find out as much as possible before mounting an expedition. The most prudent policy would be to ensure that some form of communication is undertaken with a view to gleaning as much information as possible regarding the inhabitants before attempting a planned visit.

Yet the vast distances involved in interstellar space rule out most of the usual forms of two-way communication. Question-and-answer sessions over distances between ten and several thousand light years would be meaningless for beings with life-spans similar to those on Earth. Communicative signals would be received from societies that have very likely become extinct. Suppose they, the superior civilization, send us a radio message, which takes about 100 light years to reach us, for one day. Then they wait for 100 years plus a reasonable amount of time for us to decode it and organize our response. How long should they allow for this 'reasonable amount of time'? Suppose they miss our reply? It is also likely that the generation which sent the initial signal has been replaced by one which does not wish to continue the project. But one thing is clear, communications over such a time lag cannot aspire to a conversation; at best it will be a cultural exchange. Frank Drake has pointed out that SETI scientists only aspire to a one-way system of communication, recognizing the difficulty involved in a conversation spanning millennia.

The most favourable option would be to begin with some form of recognition signal, followed by a stream of information to be slowly decoded over the next hundred years or so. Once one side realized that information was being transmitted, it could then begin transmitting information in return. This process could be initiated before the initial message had been decoded. The problem is that there are no guarantees that either stream of information will ever be understood.

Even if the time lag question was solved, there are still major barriers to a communicative exchange. With human communication there are various conventions and feedback loops which provide guidelines for the parties engaged in an exchange. For example, I speak in English but she answers in French, so I switch to French. Such loops are essential to human communication. In many cases feedback loops are extremely sophisticated, where a facial gesture or slight variation in tone can steer a conversation in an entirely different direction. Consider the difficulties involved in conducting a telephone conversation with only a rudimentary understanding of the language involved. Note how much easier it is when one can read the facial expressions and the visible responses from the other person. Without a shared background we could not even start a conversation. For example, suppose they interrupt our radio or TV programmes? How would we distinguish a message from interference? We would need the equivalent of the Rosetta Stone to decipher it. This is, of course, the philosophical problem of a communicative starting point. Appeals to the universality of science and mathematics have been made but all require some degree of tacit knowledge or prior agreement.

One solution to the joint problems of time lag and the lack of prior agreement was proposed by the Princeton physicist, R.N. Bracewell (1974), who suggested that the advanced civilization might find it expedient to send out probes to the vicinity of the planet they intend to communicate with. When the probe arrives we would have the basis for an exchange of information. On arrival the probe would be activated, first informing them of the presence of life, so that they would be on the alert for a message. But the communication problem remains: how does it attract our attention and how do we respond to it? Maybe there are probes here already, but we lack the ability even to recognize them. Perhaps, as in Kubrick's *2001*, they have left an artefact – on the Moon – such that it will not be found until we have begun space exploration and have reached a technological level which enables us to communicate. Gregory Benford (1990) suggests that an artefact may have been left on the Moon's far side, or in a stable part of the Earth's surface, which are relatively unexplored today. Yet without a common framework of discourse, even if the probe signals 'We are from outer space', we might not listen.

Bracewell's suggestion is that an intelligently directed probe would listen to our familiar broadcasts and play them back to us after a time delay, like an echo, corresponding to the time it takes for their broadcast to reach us. For example, the newscaster might say 'That is the end of the news, good night', and five minutes later it is repeated. This would certainly make us sit up. As it repeats our broadcast we repeat, the probe returns it, and so on. All that the probe would need is a receiver, an amplifier and a transmitter. Bracewell's example is based on the story of *The Count of Monte Cristo*, where the central character is imprisoned in the Château d'If. The next dungeon was 30 feet away but he did not know if it was occupied. He tapped on the stone and the other prisoner tapped back: two taps out, two taps back, and so on as conventions were built up step by step. This, of course, is possible if both share access to a similar code, such as Morse code. But what if they do not? A lot of redundant work would have to be carried out. Certainly the repetitive playback of our broadcasts would alert us to the likelihood that an intelligence was at work, but we would still need a common framework of discourse to proceed further.

Questions can be raised regarding Bracewell's proposed long-delayed echoes. Certainly they would attract attention if they disrupted radio and TV programmes, but they could be explained away in terms of natural phenomena. In the 1970s Anthony Lawson, a British electronics engineer, discovered that refractions of radio waves in the ionized region of our upper atmosphere can cause long-delayed echoes. Thus if the probe chose this potentially confusing method it would risk being dismissed as a natural phenomenon. The best thing the probe could do – in the honourable traditions of science fiction movies – is to repeatedly interrupt our broadcasts with its own message, which ought to contain a significant amount of scientific wisdom, such as a solution to a long-standing problem which is presented in a conventional manner.

In some respects, however, the communicative problem is not as difficult as it might seem. If limited to radio signals, like the prisoners tapping the wall, a great deal of background agreement will have to be employed. But this is very much a limitation imposed by the limits of present technology. Simultaneous transmissions of visual and audio data, for example, reduce the reliance upon background agreement.

The major problem with proposals for communication is the limit to background agreement. An ET civilization would have representations of their world which would reflect their systems of knowledge, as do ours. This is why a purely descriptive exchange of symbols cannot work. On Earth, when communities, or even individuals, have apparently lacked common reference points, it has been possible for them to engage in certain shared practices (work, defence, hunting, eating, procreating) which can bring different systems together. Practical activity brings us into contact with the material world and enables us to generate scientific knowledge. This is clearly missing in most schemes of communication with ETI.

The galactic network

One intriguing solution to the communication problem was proposed by Timothy Ferris (1992) who suggested that there might be a galactic network, consisting of probes distributed around the galaxy. According to Ferris, computer-controlled robot engineers can be sent to mineral-rich planets and asteroids where they will construct larger machines which will then in turn construct computer-controlled radio antennae, which can accumulate, receive and broadcast information. Further probes can be manufactured on the site by robots and sent to other mineral-rich planets or asteroids, eventually extending the network to other star systems throughout the galaxy. Communicative intelligences can hook into the network at relatively nearby stations and receive or transmit information which can be disseminated to any other intelligence within reach of the system. Although such a system is beyond the capabilities of present technology on Earth, the idea violates no known laws of physics and information technology. The digital computer is easy to replicate as it is based on the realization that anything that can be quantified can be digitized. Silicon, the basis of the microchip, is basically sand, and is widespread throughout the universe. Nevertheless, such a scheme is based on a wildly optimistic belief in technological progress and instrument reliability. It is also worth stressing that what may be *permitted* by known laws of physics is not necessarily *possible*. Theories concerning robotic probes will be critically examined in Chapter 6.

The idea of membership of a galactic network has long been one of the optimistic objectives of SETI research. Project Cyclops (1971) arrived at the idea of such a network with reference to the age of some of the earliest stars in the galaxy, which were formed over 9 billion years ago. If one takes the 4-billion-year gestation time for life on Earth as a norm, then there could have been advanced civilizations in the galaxy as long as 5 billion years ago. By now many of them will have achieved interstellar contact, and as a result will be capable of exchanging large bodies of accumulated knowledge – for example, giving information about how the universe was 5 billion years ago. This knowledge could be passed on to new races who would add to it.

A network of this kind raises several imaginative possibilities and overcomes some of the problems of interstellar communication. It would alleviate the question and answer time lag, as terminals could be located near communicating worlds. A request for information from the network could be dealt with in years, rather than millennia. Communication would not be directly with beings from other cultures, who may be hostile, and it would not matter if information came from beings whose civilizations have become extinct. Communication would be with computers rich with information derived from all over the galaxy. The network could survive the death of societies, even whole worlds, and go on accumulating information, extending its scope, and disseminating it. Hundreds, possibly thousands, of worlds could keep in touch with each other, and yet it would only require one intelligent species in the entire galaxy to initiate it.

Communication via the network need not be confined to question-and-answer sessions, but, as Ferris suggests, could also be mediated by virtual reality (VR) computer interface systems. The network could store menus of VR options whereby extraterrestrials could communicate simulations (in VR jargon, 'sims') of their world. This is not so futuristic. Although still in its infancy, cyberspace technology can provide fairly convincing 'sims' of the Martian landscape as well as a range of physical experiences, and Zoo Atlanta has a VR system wherein viewers wear a headset which transforms them into an adolescent member of a group of gorillas.

One question remains. Has such a network already been initiated by some other advanced intelligence, and if so, why has it not made contact with us? It only requires one civilization in the entire galaxy to set the network in motion, which could then continue endlessly replicating long after the demise of that society. Such a possibility would greatly enhance calculations for the factor of L – the duration of an advanced intelligent civilization – in Drake's equation (see Chapter 2). In the absence of any signal, the best that can be said is that probes may have been set up near not too distant stars in our galaxy and are sending out signals and awaiting our request for membership.

2

THE SCIENTIFIC STATUS OF
SETI

The establishment of interstellar contact may greatly prolong the life expectancy of the race that does so. Those races that have solved their
ecological and sociological problems and are therefore very long lived
may already be in mutual contact sharing an inconceivably vast pool of
knowledge. Access to this 'galactic heritage' may well prove to be the salvation of any race whose technological prowess qualifies it.

(Oliver and Billingham, 1973: 169)

Introduction

This chapter considers some of the methodological objections to SETI research
but it is maintained that, notwithstanding alleged fallacies attributed to some of
the arguments of its exponents, SETI falls within the province of genuine
scientific inquiry. SETI differs from pseudo-science in its adherence to scientific
theories and requirement to provide legitimate explanations. A review of the
post-positivist approach to the philosophy of science developed by Sir Karl
Popper, Paul K. Feyerabend, and Thomas Kuhn suggests that SETI offers a
research programme that is compatible with late-twentieth-century ideas about
science. This chapter also focuses on the underlying beliefs and assumptions
governing SETI research, and the Drake equation is examined.

SETI and pseudo-science

SETI emerged as a recognized branch of science in the early 1960s. Initially it
adopted the acronym CETI, which meant 'Communication with extraterrestrial
intelligence'. The word CETI is also the Latin term for a whale, which is an
intelligent non-human creature. More recently the expression SETI has been
adopted. SETI scientists are largely committed to radio searches as a means of
establishing communication with ET life, but their scientific theorizing is much
wider than radio astronomy.

Since 1960 radio telescopes throughout the world have sought in vain an
intelligent extraterrestrial signal. Over this period the status of SETI among the

39

scientific community has fluctuated. Its lowest point was in 1978 when Senator William Proxmire of Wisconsin presented the Golden Fleece Award to NASA's SETI scientists. This award, presented with massive media coverage, was devised by the Senator to highlight foolish pursuits which cheat the Government out of its gold. This kind of attention caused NASA to withhold funding for SETI projects, and Proxmire's amendment to the appropriation bill for NASA funding in 1981 excluded SETI research. Proxmire did, however, withdraw his amendment after a meeting with Carl Sagan, who played an influential part in persuading him of the value of SETI research, and SETI funding was restored in 1983. Recognition from the scientific community came in 1982, when the International Astronomical Union adopted the term 'Bioastronomy' to depict the search for life and intelligence in the cosmos and initiated a commission devoted to this purpose. It adopted the following seven principles:

1   To search for planets in other stellar systems.
2   To study evolution of planets and their possibilities for life.
3   To detect extraterrestrial radio signals.
4   To investigate organic molecules in the universe.
5   To detect primitive biological activity.
6   To search for signs of advanced civilizations.
7   To collaborate with other international organizations such as those devoted to biology, astronautics, etc.

This does not, of course, mean that the majority of astronomers are devoted to the objectives of SETI. There are over 7,000 professional astronomers world-wide, but less than 1 per cent of these are actively engaged in SETI. Nevertheless, it might be noted that it is SETI objectives which foster public interest in astronomy and professional astronomy is largely dependent upon the public purse. Media attention, possibly manipulated by scientific agencies, has a profound effect on the funding and direction of scientific research. Following claims in 1996 about early life on Mars, funding for prospective missions to Mars and the search for extraterrestrial life dramatically increased. Thus in 1998 estimates for the growth in NASA's astrobiology budget was predicted to grow from below US$15 million in 1999 to as high as US$100 million the following year (Seife, 1998b).

From the outset SETI scientists have sought to distance themselves from Ufology and other branches of inquiry deemed to be pseudo-scientific. There are clear methodological rules regarding the recognition of ETI: for example, the rule that every object must be assumed to be of natural origin unless it is decidedly proven to be unnatural, places the burden of proof on SETI researchers to provide adequate evidence for their theories. This is in sharp contrast to many exponents of pseudo-science where the onus of proof is directed to the sceptic. SETI scientists work within the limits of existing theory and levels of technology and eschew appeals to as yet unheard of theories when offered as solutions to today's insoluble problems. The case for SETI, it must be stressed, rests on its appeal to plausibility which is based on knowledge derived from state-of-the-art technology. Confidence in SETI's chance of success is predicated on developments in astronomy, space exploration, information technology, and the life sciences which offer a sound empirical foundation for SETI research.

There are, however, several logical pitfalls in the arguments which frequently appear in SETI literature; and philosophers of science have drawn attention to them. Ernan McMullan (1971) suggests that two types of probability may be sometimes confused in arguments concerning the likelihood of extraterrestrial life. The first type is inductive probability which is based upon a frequency count of outcomes: this might involve a count of heads or tails in a coin-tossing experiment or predictions based on a correlation between traffic accidents and teenage motorists. Conclusions drawn from inductive probability require neither an understanding nor an explanation of the processes involved. Thus a correlation may be made between teenage motorists and traffic accidents without recourse to any explanations why teenagers are involved in more or less accidents than other motorists. The second type of probability, theoretical probability, does require at least a partial understanding of the causes. Thus if one has a rea-sonable theoretical explanation of the psychological factors affecting teenage motorists, one could attempt a calculation of the percentage of likely accidents without observing numerous instances. Now appeals to inductive probability are of little value to SETI as we only know of one planet where intelligent life has evolved. But appeals to theoretical probability are not without methodological problems: although theoretical progress is underway, we are not yet in a position to advance a *definitive* theory as to how planets, life and then intelligence, came about. This lack of a firm theoretical base, argues McMullan (1971), is fundamentally damaging to the SETI enterprise and partly accounts for four fallacies which he detects in SETI literature.

The first fallacy, according to McMullan, is the belief that given enough time 'the probability of any natural outcome dependent on universal natural processes increases to virtual certainty' (ibid.: 292). For example, given enough time, any environment containing the constituents associated with life will inevitably produce life. The problem with this argument is that, lacking a theory of how life came about, it appeals to nothing more than a random juxtaposition of elements. But this is unlikely, as what we do know about life is that it emerges out of a gradual biochemical process involving, as McMullan says, 'a host of unknown interdependent environmental factors' (ibid.). The problem is not resolved by arbitrarily lengthening the timescale, but rather by knowing which of the essential factors will be present and what forces will be operating upon them.

The second fallacy McMullan identifies is the 'uniformitarian fallacy'. This fallacy is found in the argument which states that if life can develop in one place it must inevitably develop in another place with similar conditions. This argument is sometimes linked with appeals to general scientific beliefs such as the belief that nature does not admit uniqueness. Again, McMullan identifies the fallacy with reference to our lack of a theory as to how life came about: for unless we have this theory for one environment we can have no idea how life will emerge elsewhere. The same appeal to the lack of a theoretical account of the emergence of life underpins McMullan's depiction of the third SETI fallacy, the appeal to large numbers, the vast numbers of galaxies, stars and possible life-sustaining planets. Unless we have a theory of some sort we cannot attach any theoretical probability to any predicted outcome. It might be said that it is logically possible that life can emerge elsewhere, but expressions like 'logically possible' cannot be assigned a numerical rating if there is no theory to support it.

It might be countered that evolutionary theory provides a theoretical grounding for speculations about extraterrestrial life. But McMullan cites SETI's appeal to evolutionary theory as an example of the fourth fallacy; namely that it is fallacious to use the theory of evolution as a predictive theory. SETI scientists often argue that once life originates the operation of natural selection will inevitably lead to more complex life-forms and ultimately to consciousness and intelligence. Yet biologists constantly point out that no predictions can be made on the basis of evolutionary theory; intelligence may bestow selective advantage, but its emergence cannot be guaranteed as other features may turn out to be more advantageous.

How damaging are these alleged fallacies in the reasoning of SETI exponents? They only appear to be damaging if the arguments which contain them are employed to support an *a priori* claim that ETI exists. If, however, they are used in the context of discovery, as steps in a campaign for exploration, as reasons towards the plausibility of conducting an empirical investigation, they cannot be faulted. A rebuttal of the four 'fallacies' might read as follows:

1   A useful realistic strategy could rest on the belief that the chance of success is increased with an increase in the time available.
2   Uniformitarianism is only fallacious if one insists that uniform conditions must yield uniform results without exceptions; in many cases it is a good rule of thumb to seek similar results from similar conditions.
3   Likewise, an appeal to large numbers, while providing no guarantee for any particular prediction, is suggestive that a wider search may be productive.
4   Moreover, if claims regarding the inevitable emergence of intelligence are replaced with suggestions that intelligence is a likely development in some life-forms, for which it is selectively advantageous, then no harm is done to the non-predictive status of evolutionary theory.

It is important, therefore, to disentangle arguments which indicate that an empirical search might be fruitful from arguments which attempt to establish the existence of ETI on *a priori* grounds.

SETI and the problem of scientific methodology

Questions regarding the scientific status of SETI inevitably give rise to further questions concerning the very nature of science itself. The traditional or received view during the first half of the twentieth century was that of logical empiricism, according to which all scientific theory must be grounded in observational experience. A fundamental distinction was drawn between observational terms and theoretical terms, and a central part of the logical empiricist programme was given to the attempt to show how theoretical terms could be interpreted on the basis of observational experience. The meaning of scientific statements would be examined with reference to observations which provided verification. Observational discourse was taken to raise no problems, as it referred directly to experience. Another central feature of the received view was that the history of science exhibited a pattern of progress, recording the gradual removal of superstition, prejudice and erroneous theories.

During the latter half of the twentieth century the received view of science came under attack from numerous sources, among whom were Sir Karl Popper, Thomas Kuhn and Paul K. Feyerabend. While the fundamental differences between the three philosophers continue to be debated among philosophers of science, the damage they inflicted upon the received view of scientific methodology would seem to be irreparable. Popper demolished the idea that empirical verification was the essence of scientific methodology and maintained that the most important characteristic of honest science was the potential for hypotheses or conjectures to succumb to falsification.

Popper's doctrine of falsification provided him with a criterion of demarcation between genuine science and pseudo-science. Valid scientific theories are those subjected to, and able to withstand, the test of falsifiability until they are eventually overthrown. Scientific theories were not ultimately reducible to foundational observational experiences and they cannot ever be accepted as final truths; they are conjectures awaiting refutation. Thus the function of the theoretician is to propose scientific conjectures, while the function of the experimenter is to devise every possible way of falsifying these theoretical hypotheses. The confirmation of a hypothesis, maintained Popper, has no scientific meaning.

Popper is one of the few among modern philosophers of science to have influenced the work of scientists, and his appeal to potential falsification captures those moments in science where research efforts are devoted to demonstrating the falsity of particular claims. This, of course, is relevant to our understanding of the status of SETI. Scientists engaged in SETI are unlikely to cling tenaciously to hypotheses concerning colonies of extraterrestrials on Mars once overwhelming evidence from space probes provides falsification. Only members of cults and various fringe activities will go on to produce exotic reasons for maintaining their beliefs against overwhelming contrary evidence. It will be maintained throughout this discussion that, as a branch of scientific research, SETI conforms to Popperian criteria for demarcation.

Further objections to the received view of logical empiricism were developed by Paul K. Feyerabend and Thomas Kuhn in the 1960s, and these continue to generate controversy (Munévar *et al.*, 2000). In a series of polemical essays Feyerabend (1975) not only rejected the received view of logical empiricism and went beyond Popper's critical rationalism; he actually attempted a demolition of all attempts to construct a methodology of science. Opposing the very idea of scientific methodology, Feyerabend drew a sharp contrast between the actual history of science and the so-called logic of science, arguing that the most successful scientific research programmes have never proceeded according to rational methodologies at all. He introduced a slogan 'anything goes', and in detailed studies of Galileo's defence of Copernicanism argued that the ultimate success of Galileo's research programme lay not in any superior scientific method, but in a variety of ways whereby opponents were disarmed and supporters were convinced that they were on the right track. Like many great thinkers, Feyerabend was to revise his position throughout his life, but his fundamental message remained intact: namely, that there are numerous examples throughout the history of science where embryonic research programmes, however much they depart from established theories and facts, have revealed an ability to flourish, and add to the store of human knowledge, and possibly overturn the existing orthodoxy. SETI, so it would seem, despite charges that it is a science without a subject-matter, is very much akin to those research programmes which, according to Feyerabend, require breathing space and the relaxation of criteria for rational assessment.

Thomas Kuhn was a contemporary of Feyerabend, who shared the view that the actual history of science did not fit well with the normative image of science that was projected by the logical empiricists. In *The Structure of Scientific Revolutions* (1962) Kuhn appealed to case studies in the history of science in order to establish the view that scientific development was more dependent upon a set of shared presuppositions, practices, values and beliefs, held within a community of scientists, than on the accumulation of facts upon observational foundations, as the logical empiricists had maintained. Initially, Kuhn adopted the term 'paradigm' when referring to a particular conceptual framework within which certain facts and theories were meaningful. Great scientific revolutions, he argued, were akin to political revolutions whereby the existing orthodoxy, when confronted with anomalies, degenerated into a state of crisis before being replaced by the new order. A major transformation in scientific thought, he held, is not so much brought about by the falsification of the old theoretical framework or the verification of the new. Rather, the transformation (which became known as a 'paradigm switch') involved a change in both the background theory and the observations bound up with it. In this way Kuhn replaced the linear notion of scientific progress with the idea of periodic conceptual transformations or changes in the way that science is actually conducted. A scientific revolution, Kuhn argued, does not lead to a better or worse standpoint, as the meaning of the terms within one paradigm may be incommensurable with terms employed in another paradigm. Obviously, there may be benefits associated with the new paradigm – post-Copernican astronomy solved the problem of predicting planetary motion – but there may be losses involved with a scientific revolution, which are termed 'Kuhn-loss'. For example, the consolidation of science-based medicine brought major benefits in mortality and morbidity rates but Kuhn-loss can be discerned in the widely forgotten practices of natural medicine.

Early responses to both Feyerabend and Kuhn were hostile, with accusations of anarchism and mob-psychology levelled at both of them. Feyerabend's notorious slogan, 'anything goes', was seen as a way of endorsing witchcraft, astrology, Ufology, and many other areas of inquiry traditionally reserved for cranks. Kuhn's appeal to paradigms as frameworks within which particular forms of scientific discourse are meaningful was criticized (and even sometimes defended) as a means of permitting all sorts of fringe activities to claim scientific status, as long as they could indicate that their activities were meaningful within a paradigm. Nevertheless, despite these objections, the legacy of Feyerabend and Kuhn is their demonstration that an excessive concern with the methodology of science is incompatible with developments in frontier areas of scientific inquiry.

A programme that is suggestive of solutions to long-standing problems will nevertheless have to survive the scrutiny of the wider scientific community, but it is unlikely to be dismissed with an appeal to the canons of scientific method.

Is SETI scientific? The answer is yes, if we interpret science as something that scientists do. Opponents have maintained that it is unscientific, a form of pseudo-science, but this label only applies to some of the fringe activities which pay little attention to scientific facts and theories. Pseudo-science is attached to most branches of science where there is a broad area of public interest. The medical sciences are surrounded by a vast array of pseudo-sciences which include crackpots. SETI likewise reaches to the fringes but nevertheless adheres to established scientific facts and theories. It does, however, differ from much of scientific activity in the sense that it is driven by metaphysical beliefs and deep psychological desires for companionship. But world-views do not have to be antithetical to scientific evidence, and metaphysical and psychological factors are a long-standing respectable feature of good science, which were only temporarily excluded in the heyday of logical empiricism.

SETI is also distinguished from fringe science by its acceptance of Occam's razor, the imperative to limit explanations to the plausible and testable before admitting more exotic hypotheses. This is well reflected in the ongoing and seemingly endless controversies over Martian life. Following the indeterminate findings of the Viking experiments, N.H. Horrowitz said:

> It is impossible to prove that any of the reactions detected by the Viking instruments were not biological in origin. It is equally impossible to prove from any result of the Viking experiment that the rocks seen at

the landing sites are not living organisms that happen to look like rocks. Once one abandons Occam's razor the field opens to every fantasy.

(1977: 61)

SETI is still a young science and therefore it can be argued, in accord with Feyerabend's account of scientific research, that it should be given enough breathing space to grow. But how does it compare with other young sciences? It is approximately the same age as molecular biology and older than chaos theory. It differs fundamentally from these two because its subject matter might not even exist. On the other hand, just one authentic detection would yield massive rewards and a place in posterity alongside Columbus, Galileo and Darwin. This is why it is attractive to many young and talented scientists. SETI has also focused attention on what we mean by 'intelligence' in a way that may turn out to be more philosophically rewarding than inquiries into artificial intelligence systems. It offers a cosmic standpoint for the understanding of intelligence and forces a critical assessment of hidden assumptions concerning its origin and function.

SETI's richness may encourage suggestions that its claims are non-falsifiable. If not on the Moon, then extraterrestrial intelligence is to be found on Mars. If not in the solar system, then other stellar systems, somewhere in the galaxy or other galaxies. This is not to say that SETI scientists tenaciously cling to refuted theories, but simply that even when refuted, hypotheses concerning the possibility of life in certain sites, can be re-applied to other sites. The avoidance of falsification is not a ruse to protect a research programme, but simply reflects the fact that only a small fraction of space has been searched. Although, strictly speaking, the hypothesis that there is extraterrestrial life cannot be falsified, SETI scientists base their research on empirical statements and observations which can be and are falsified. Moreover, it is unscientific to conclude that an ETI hypothesis has been tested and failed unless a really extensive search has been carried out. There are also strict rules governing confirmation, which in SETI research is not vacuous as in spiritualism where almost any experience may be deemed confirmatory. SETI research has applied strict regulations over confirmability, such that no purported signal has yet had an acknowledged confirmation. Repeated failure, however, might not be seen as decisive falsification, but it would certainly contribute to a loss of interest and a reduction in resources.

SETI differs from Ufology and branches of fringe science by its adherence to its methodology of scepticism, requirement for verification, and acceptance of the parameters of scientific knowledge. Its maturity is reflected in its reluctance to support wild claims or endorse inadequately documented data. Its researchers accept that it is normal for the propounder of an exotic hypothesis to provide argument and evidence to convince the sceptic. A failure to do this has dogged Ufology from its beginnings. Nevertheless, as in all branches of science, it is not the case that scepticism must rule in the absence of overwhelming proof, and acceptable means of neutralizing the sceptic can be invoked. Carl Sagan (1983) thus appeals to the history of deprovincialization of our world-view since the sixteenth century in this context. This history reveals the reluctant acceptance of the fact that our Sun is merely one of 400 billion stars, that there is nothing special about us or our planet, and that we emerged along a common evolutionary process with other animals. The history of deprovincialization cannot be appealed to as proof that ETI exist, but, says Sagan, it urges caution in accepting the sceptic's standpoint. It is worth noting that, while Carl Sagan's appeal to the assumption of mediocrity is a valid counter to the assumption of uniqueness, neither assumption can generate its own truth. Ultimately, however, 'the only valid approach to the question is experimental' (Sagan, 1983: 113).

Is SETI a paradigm in the Kuhnian sense? Apart from the optimism of its exponents it is not committed to any revolutionary epistemological rupture with the past nor does it reflect a crisis in any contemporary branch of science. Yet it does exhibit a set of ground rules and doctrines characteristic of a Kuhnian paradigm. Its scientific credentials are underpinned by three fundamental features of the universe. First, the universe is believed to be uniform with the same essential building materials, the same chemical elements, functioning according to universal natural laws. In fact one essential question which SETI has raised concerns the universality of biological laws. Kepler, Galileo, Newton and Einstein postulated physical laws which dominated the universe. The question at stake in the extraterrestrial life debate is whether there are similar universal biological laws. The basic chemicals of life are widely distributed throughout the universe. As noted, the appeal to uniformitarianism is only fallacious if it is claimed that similar consequences *must* follow from similar conditions. The assumption that the universe is homogenous and isotropic also simplifies models used to describe it.

The second feature of the universe is supported by two major revolutions affecting modern science: the Copernican revolution revealed that we are not privileged in our location and the Darwinian revolution showed that we are not a privileged species. There is no unique spot, or centre, in the universe; there is nothing unique about our Sun, and no reason to assume that events on our planet could not happen elsewhere. In 1996 the Hubble telescope provided one of the deepest-ever views of the universe, forcibly demonstrating that the Milky Way Galaxy is only one of billions of galaxies in the universe, and is not even centrally located. Modern cosmology supports SETI's plausibility: the universe has no centre, no outer limit or surface, no privileged place, but continuous expansion. Further developments in modern cosmology include speculations that there are many universes, in which case it cannot be claimed that the universe itself has a privileged status!

Third, the universe is massive: within telescopic range are over a hundred billion galaxies, each containing hundreds of billions of stars. Or, as it is often said, there are more stars than there are grains of sand on all the beaches on Earth. It must, however, be stressed that the appeal to these three features plays a heuristic rather than a probative role in SETI research; there is no substitute for an empirical investigation.

In addition, SETI researchers are committed to two basic beliefs. The first is the belief that intelligent alien life-forms will have similar thought patterns to those displayed by intelligent life on Earth. In this respect SETI has been described as a search for extraterrestrial human intelligence (SETHI). This does not have to commit SETI scientists to a search for other human beings. Carl Sagan has argued that, given different evolutionary programmes and different environments, it would be astonishing if ET life resembled ours. But while biological diversity is expected, there is a belief among SETI scientists in a similar technology. This is because of the universality of physical laws. Thus the authors of the Project Cyclops Report state that:

> Regardless of the morphology of other intelligent beings, their micro-scopes, telescopes, communication systems, and power plants must have been at some time in their history, almost indistinguishable in working principles from ours. To be sure there will be differences in the order of invention and application of techniques and machines, but technologi-cal systems are shaped more by the physical laws of optics, thermody-namics, electromagnetics, or atomic reactions on which they are based, than by the nature of beings that design them.
>
> (Oliver and Billingham, 1973: 4)

SETI scientists are not committed to the view that *every* sequence in the evolution of life towards a technological culture has to be replicated. This would be impossible. What matters is that, by one way or another, intelligent extra-terrestrials develop a technological ability similar to ours. This may involve many alternative sequences – as in the evolution of science on Earth – and many different ways of arriving at the same goal, given broadly similar problems. If physics is universal, they will make use of the same laws, although discoveries and applications associated with them will be diverse. They may discover how to harness the power of electricity in ways that are not dissimilar to ours, but it would be very unlikely for them to invent electric toasters or battery-operated door chimes. With a broad agreement over the principles of science, it is presumed that there would be few problems in constructing a dialogue based upon technology, science and mathematics. This presumption is not without its critics (Weston, 1988).

The second belief is that communication with other life-forms will be via electromagnetic signals at some 'magic' frequency that corresponds with fundamental properties of atoms, such as hydrogen, which are commonplace throughout the universe. A further aspect of the SETI paradigm, which shall be examined here, are the assumptions and doctrines built around the Drake equation. The main components of the equation will be briefly examined here, but a more detailed discussion of its component theories of planetary formation, the evolution of life and intelligence, will be covered in the next chapter.

The Drake equation

The Drake equation was first presented by Frank Drake at a meeting of scientists gathered to discuss 'Intelligent Extraterrestrial Life' at the Green Bank Observatory under the auspices of the Space Science Board of the US National Academy of Science in November 1961. The equation has consequently become a SETI creed, its logical justification. Reflecting on the equation some thirty years later, Drake said that its basic premise 'is that what happens here will happen with a large fraction of the stars as they are created, one after another, in the Milky Way Galaxy and other galaxies' (Drake, 1990: 151). This is the equation:

$$N = R* \times f^p \times n^e \times f^l \times f^i \times f^c \times L$$

It is not a definitive equation, not is it inviolable like $E = MC^2$. Its components can be broken down as follows:

N = the number of technically advanced civilizations in the galaxy that are currently capable of communicating with other solar systems.

$R*$ = the number of new stars formed in the galaxy each year.

f^p = the fraction of those stars that have planetary systems.

n^e = the average number of planets in each such system that can support life.

f^l = the fraction of such planets on which life actually exists.

f^i = the fraction of life-sustaining planets on which intelligent life evolves.

f^c = the fraction of intelligent life-bearing planets on which beings develop the means and the will to communicate over interstellar distances.

L = the average lifetime of such a technological civilization.

R*: the rate of star formation

There are held to be over 40 billion stars in the Milky Way Galaxy with enough loose gas and dust for millions more. Estimates put the rate of star formation at ten each year. This, however, is a theory-laden guess in that any attempt to assess the rate of star formation must be derived from a theory as to how stars are formed. The dominant theory is the Big Bang theory, the moment when the universe began. According to this theory the first stage was when the elements flew off in rotating clouds. Then they cooled and formed galaxies. The early stars were massive combinations of hydrogen and helium, and when they exploded heavy elements were created and flung about the universe. Eventually other generations of stars condensed from this matter, leaving their rocky substances that formed planets in their orbiting paths. There are countless of billions of stars in over a billion galaxies. A star's colour is a clue to its surface

temperature and the scale for classifying stars is indicated by the letters O B A F
G K M, running from the hot blue stars to the cool red ones. A star's life-span is
measured in millions of years. Our Sun, for example, is intermediate and is
classified as a G type. It has existed for about 5 billion years and supports life on
Earth. It is likely to survive for another 8 billion years. Similar stars to our Sun
can be found in the F to K groups. Whether they can support solar systems
similar to our own is a matter for speculation. Application of the Goldilocks
criterion eliminates the remaining groups as either too hot or too cold. It is
estimated that about 25 per cent of the total number of stars are found in the
intermediate F–K groups. From this it is concluded that a quarter of all the stars
in the heavens are capable of supporting planetary systems with life.

The problem with this kind of reasoning is obvious. The figure of 25 per cent
is only a theoretical possibility, not an inductive probability. The data available
from astronomers concerning the existence of planetary systems are far too
limited for the assigning of inductive probabilities. The potential for confusion
between theoretical possibility and inductive probability re-emerges throughout
the whole debate on ET life.

f^p: the fraction of stars with planets

Excluded here are classes of stars unlikely to have satellites, such as some, but
not all, of the double stars and stars with short life-spans. This, of course, does
not provide any degree of certainty concerning the numbers of planets to be
found in the vicinity of other types of stars. It is a theory-laden estimate, based
on theories of planetary formation and indirect observation of the alleged effects
of planets on their parent stars. When Drake formulated his equation no planets
had been directly observed outside of the solar system, but during the past few
years there have been over thirty claims identified as planets.

The belief that planetary systems are the rule rather than the exception is
supported by current astronomical theory. Estimates concerning the number of
stars with planets will vary according to which theory of planetary formation is
adopted. The Catastrophe theory of the nineteenth and early twentieth centuries held that planets were the product of the debris from either collisions or the
explosions of stars. Given the vast distances between stars and their durability
this would suggest that planetary formation was infrequent. Catastrophe theory
thus dramatically reduces the theoretical scope for f^p. However, during the past
thirty to forty years the nebular theory of Descartes, Kant and Laplace has re-
emerged. The current interpretation of this theory maintains that planets are the
offshoots of the large rotating clouds of dust, out of which the stars were
formed. Accordingly, planets are very likely companions of stars as the theory
maintains that they are products of the same creative forces. This is described as
accretion by impact. Dust grains stick together, welding on impact. Once the
planetesimals are large enough, their gravity will attract other particles. The
most prevalent model suggests two things: first, that planetary systems are

common, and second, that our solar system is typical. According to Bruce Jakosky (1998: 238), computer simulation of planetary formation indicates that 'Terrestrial-like planets appear to be a natural consequence or by-product of the formation of stars, and they should be widespread throughout the Galaxy.'

To date, about one quarter of the young stars that can be observed have protoplanetary disks around them. On these terms we can expect that ex-solar planets will be widespread. If stars and planets are thus formed together, at the rate of 10 planets for each star, there would be a minimum of 400 billion planets in the galaxy. Nevertheless, f^p is still very much in the realm of theory; there is no compelling proof that many of the rotating clouds have evolved into planets. Confirmatory empirical evidence would appear to be moving in support of the belief that planets are the rule, rather than the exception (see Chapter 3). Studies of young stars (about 1 million years old) in the Orion Nebula have been undertaken by means of the Hubble telescope. Photographs have revealed a number of dusty disks which, according to the prevailing theories of planetary formation, could become planetary systems in the Orion Nebula, so it would appear protoplanetary disks are the rule. Photographs from the Hubble telescope also support theories regarding the stability of protoplanetary disks, and hence, theories which suggest that planetary systems are formed by the slow accretion of dust particles which stick to each other for over millions of years until they become planets (Reichhardt, 1996).

Although searches for ex-solar planets are usually conducted in the vicinity of stars, which conforms with the standard model of planetary formation, there have been several claims regarding the possibility that there are millions of nomadic planets in the Milky Way Galaxy alone (Muir, 2000: 14). A British team of astronomers, using the infrared telescope on Mount Kea, Hawaii, discovered quite a number of objects about five to thirteen times as heavy as Jupiter which do not orbit a star. This survey confirmed similar results by a Japanese team of astronomers in 1998. Although these claims might push up the number of planets in the universe, current theory suggests that these nomadic gas giants, with low levels of carbon and oxygen, are unlikely to be inhabited.

n^e: the number of planets that could support life

This, again, is an appeal to theoretical possibility. Chemical surveys suggest that there is nothing unique about Earth material and that all the matter in the solar system and beyond has a common origin. By extension of the theory the same building blocks which are necessary for life will be found throughout the universe. This, however, only specifies the necessary, not the sufficient, conditions for life, as other intervening factors may prevent the actual emergence of life. The essential chemicals for living structures may be in abundance but the environment might not be favourable. It would require a planet with liquid water; the presence of elements necessary for life to use either in metabolism or reproduction; a source of energy that is available to biota, and an environment that is sufficiently stable for biota to survive and evolve. Extreme temperatures would inhibit many forms of life, although in the past few years several micro-organisms have been discovered on Earth which can exist in extremely inhospitable conditions. Nevertheless, most forms of life will require a stable temperature: a planet too far away from the Sun, like Mars, will freeze in a stillborn death. If it is too close, like Venus, it will suffer from a runaway greenhouse effect. Strictly speaking, n^e is limited to the number of planets which are found in a habitable zone. Even on the most optimistic estimates the number assigned to n^e would be restricted to planets with an orbit similar to that of the Earth around the Sun, with a similar ecosphere capable of sustaining life, and even then the estimate would remain a theoretical possibility. Appeals are sometimes made to a 'habitable zone' within the Milky Way Galaxy, which is some 1,500 light years wide and 30,000 light years from the centre, which contains over a billion stars, many of which could provide habitable stellar ecospheres. According to Oliver and Billingham (1973), the authors of Project Cyclops, there are probably in the order of 10^{10} potentially habitable planets in the galaxy.

In any discussion regarding potential sites where life *could* develop, it is important to resist extreme forms of biological determinism, which entails the view that given the right conditions life will inevitably emerge. This belief is not grounded in any known scientific laws. The point about searching for planets where life could develop, and the attendant analogies with life on Earth, is that such sites are more plausible candidates for the search. In this respect discoveries of potentially habitable sites are supportive of a campaign for exploration, not as evidence of possible life.

Nevertheless, a significant increase in the number of potential sites might be derived from a consideration of satellites of large gas giants like Jupiter, which generates a considerable source of heat and could be regarded as a mini solar system. There may also be numerous nomadic dark planets derived from matter which has been expelled from a stellar system in its early stages of development. For example, when a new planetary system forms, the gravitational effects of the large planets could catapult Earth-sized planets out of the system. If they have internal sources of heat and hydrogen-rich atmospheres, they could produce life or sustain life that has developed elsewhere. Late-twentieth-century discoveries of terrestrial forms of life deep underground or beneath frozen lakes revealed that life can exist without sunlight. Of course the detection of dark planets would be difficult, and the only conceivable means would be to observe any dimming of light should they pass in front of a star.

f^l: the number of planets that have actually developed life

So far the factors in the Drake equation call for guesses, but appeals to fairly well-established theories provide them with a degree of plausibility. More work may be undertaken to strengthen the theory, more reliable data may contribute to the weight of empirical evidence. But from this point onwards the equation moves into the realm of conjecture. Estimates regarding the size of f^1 have no empirical support and rest on a widely disputed theory of how life developed. At present biologists have not reached a consensus regarding explanations as to how terrestrial life originated, which renders Earth an unsatisfactory model for the rest of the universe. The dominant theories refer to the evolution of organic compounds out of the oceans, when the Earth's atmosphere consisted primarily of hydrogen, nitrogen, carbon dioxide, methane and water vapour. It is argued that the action of the Sun, together with volcanic activity on Earth, was primarily causative of the emergence of organic compounds (such as the amino acids) out of inorganic molecules. If correct, this explanation provides the necessary, but not the sufficient, conditions for life. All we have is a model of the possible emergence of life; there is no guarantee that life must emerge. The theory does not show how matter came to life; this problem has yet to be resolved. As yet life has not been produced in a laboratory. Until we know much more about the origins of terrestrial life the Earth cannot be regarded as a reliable model for the emergence of life on other planets.

f^i: the number of planets with intelligent life

Conjectures regarding the number of life-bearing planets with intelligent life are predicated on widely disputed concepts of intelligence. Philosophers, psychologists, anthropologists and sociologists have failed to produce an authoritative definition of intelligence which can be applied to life-forms on Earth. There is no satisfactory definition of intelligence for either humans or other species. There are disputes whether intelligence can be attributed to non-humans, such as chimpanzees and dolphins, and whether machines can exhibit forms of intelligent behaviour. We know very little about the origin of human intelligence, of the processes and mechanisms by which it is formed. Measuring the size of brains is far from satisfactory as a means of detecting or examining intelligence, as many terrestrial species with large brains do not appear to exhibit intelligence but use their large brains to control the basic functioning of their bodies. In the case of humans, intelligence appeared to occur with developments in manual dexterity.

It should, however, be admitted that the nature of the search for ETI will itself determine the parameters of intelligent life. It is of little use speculating on the likely existence of forms of intelligence that cannot be detected by current instruments. For this reason Jill Tarter points out that the search for intelligent life is a search for technology that is detectable by our technology. She offers a more pragmatic definition of intelligence: 'A species' ability to technologically modify its local environment in ways that can be detected over interstellar distances' (Tarter, 1995: 9). This, she considers to be preferable to the 'overly complex and convoluted definition of "intelligence" offered by researchers in other fields' (ibid.).

In what might be described as a 'leap of faith', some SETI exobiologists have maintained that life, as it evolves, is likely to evolve greater forms of intelligence; that intelligence must necessarily follow life. The belief is that intelligence will enhance survival. This is not always the case, as many non-intelligent species, bacteria, and plants, have an excellent survival record. A study of terrestrial natural history is unlikely to support the belief that intelligence enhances survival, and even if an evolution towards greater intelligence does occur, we are still left with the problem of determining the nature of intelligence. SETI is not helped with speculations concerning intelligence other than a criterion for intelligence which is bound up with the emergence of a technological civilization capable of radio communication.

It may be easier to detect life than intelligent life. NASA already has a project called TOPS (Towards Other Planetary Systems) which is based on techniques for carrying out spectroscopic examinations of distant planetary atmospheres, seeking evidence that a planet's chemistry has been altered by metabolism. Evidence of a large amount of oxygen may indicate the presence of biochemical life.

Recognition that life may be more prevalent in the universe has followed a number of claims that evidence of simple organic life has been detected in Martian meteorites. Sceptics who would generally oppose SETI's belief in the prevalence of life might, however, concede that simple organic structures are more widespread than previously held, but nevertheless resist any move from f^l to f^i. Biochemists would now concede that single cellular organisms may be quite prevalent in conditions throughout the universe which resemble those on Earth some 4 billion years ago. But, having made this concession, the sceptic might insist that later stages, f^i and f^c are improbable. The argument is that while bacteria appeared fairly rapidly on Earth, it nevertheless took over 3 billion years for multicellular life to develop. This appears to support the suggestion that the transition from simple life to the more 'advanced' multicellular organisms, and then to technological civilizations, is extremely rare.

f^c: the number of planets where intelligent life has the capacity to communicate with other planets

Conjectures regarding the fraction f^c frequently rest on the belief that a scientific civilization like ours is an inevitable consequence of intelligent life. This presupposes that other evolving species are likely to pass through similar stages from the Stone Age to the Nuclear Age. But if we consider the contingencies in the development of human natural and social history such a repetition appears very unlikely. Mammals might not have survived long enough to develop a human species. Natural disasters of the kind that wiped out the dinosaurs could have arrested human development, allowing some entirely different dominant species to evolve. Humans might not have developed sophisticated coordination between hands and eyes, which is necessary for tool-making and hunting and eventual control over the environment, the minimum conditions for a scientific culture.

Intelligent human civilizations, such as those in Ancient Greece and Ancient China, flourished and died without any likelihood of discovering Maxwell's laws of electro-magnetism: the minimum condition for radio communication. A scientific civilization also requires a few lucky breaks from the physical environment. Despite their acknowledged intelligence and communicative skills, dolphins will never invent printing or acquire an adequate knowledge of electro-magnetism as a precursor to radio contact. But even if an ET civilization did develop an understanding of the laws of physics, that would still leave open the intriguing philosophical question of whether they mean the same to them as they do to us.

Generous estimates of the number of planets with intelligent communicative life suffered a serious setback in 1992 following the completion of a radio search conducted by D.G. Blair at the University of Western Australia. Blair led a team of astronomers, using a Parkes radio telescope, at the 'magic' frequency of 4.462336275 GHz during two observations in 1991 and 1992. This frequency was selected by multiplying the frequency of hydrogen by pi, the first fundamental constant considered likely to be discovered by any intelligent civilization. The search covered the neighbourhoods of 176 stars in the F, G and K range, within forty light years of the Earth. No signal was detected. These negative results weaken Drake's assumption that technological intelligence will inevitably emerge given enough time on an Earth-size planet near a Sun-like star (Blair *et al.*, 1992).

L: the lifetime of an advanced technological civilization capable of interstellar communication

Estimates for this factor have neither theoretical nor empirical support. The only advanced technological society we have knowledge of is our own and it will only be after its extinction that an accurate figure for L for one civilization can be supplied. But to appreciate how consideration of L might reduce the size of N (the number of civilizations currently capable of communicating), it is worth reflecting that it took about 300 million years of the cool planet period for life to evolve and a further 3,500 million years for only one species, out of billions, to invent the radio. And within half a century after that, intelligent life on Earth had produced nuclear weapons and began to pose an ecological threat to the planet, both of which are capable of terminating life within a very short period. It may be a mere fluke that humans developed the potential for interstellar communication before self-destruction.

Reflections on the quantification of L are bound up with a related terrestrial question: can we survive advanced technology? In fact, it was reflection upon L which enabled Carl Sagan to persuade Senator Proxmire that SETI might provide a means of understanding how our society might survive the nuclear age, as L indicates civilizations which have avoided self-destruction. It may be that L is little over a hundred years. It would be a matter of cosmic irony if civilizations turn out to be capable of destroying themselves at the very moment they achieve the ability to communicate with each other. Apart from war and ecological disaster, L can be further reduced by collisions with asteroids, exhaustion of natural resources, over-population, genetic degeneration and loss of interest in science, where people turn to religions like Zen Buddhism, new age cults or simply lapse into hedonism.

A pessimistic calculation for L was made by J. Richard Gott III (1993) who estimated an 0.2 million to 8 million years limit to longevity of our species at 95 per cent confidence level. According to Gott, the average longevity for most species is between 1 million and 11 million years, and for mammals it is about 2 million years. Gott insists that intelligence – defined in terms of the possession of self-consciousness and cognitive skills which enable abstract thought, creativity and an ability to think about the future – confers no survival value. He points out that our early ancestor, *Homo erectus*, lasted 1.4 million years and the Neanderthals lasted 200,000 years. Taking the Copernican principle – there is nothing unique or privileged about the human race – at face value, Gott argues that there is unlikely to be sufficient time for the human race to either colonize space or establish widespread contact through radio communication. Space travel, he suggests, somewhat pessimistically, has run its course. The Cold War, which supported the space race, is over. Moon travel lasted a mere four years. Human activity in space is confined largely to a close-to-Earth orbit. With increasing pollution of radio space and an energy crisis, our capacity for radio communication and space travel will soon expire. The space window may soon be closed, and will only remain open for a limited period before energy supplies fail, space debris (see Simpson, 1994), or light and electromagnetic pollution (McNally, 1994) close it.

It could transpire that the limit to the communicative abilities of a technological civilization is less dramatic than destruction by war or famine; these limits may be determined by its tendency to surround itself with a barrier of pollutants. Thus in 1993 Gott estimated that we only have the space window for another thirty-two years. If we do not use that window we shall stay and become extinct, but if we do use it we have a slim chance of survival via the spread of colonies. But he warns that our chances are slender.

With luck, perhaps, humankind will survive and remain capable of sending messages which can be picked up if there is anyone out there to listen. The only incontestable fact in this mass of speculation is that so far the human race has survived.

The belt of life and the sleeping dragon

Severe limitations upon the first and last factors in the Drake equation have been set by invoking the notions of a belt of life and a sleeping dragon at the centre of the universe. The belt of life hypothesis (Marochnik and Mukhin, 1988: 49–59) suggests a narrow circular zone passing between the spiral arms of the galaxy, Perseus and Sagittarius, which includes the Sun's galactocentric orbit. Once a planet enters these arms, all life will perish due to the radiation from the exploding supernovae there. The spiral arms, so the hypothesis goes, are places of life and death. If the galactic belt hypothesis is correct, then proximity to the spiral arms limits the first and last factor in the Drake equation: the number of suitable stars will be limited to those in the belt, not the galaxy as a whole, and the length of technological civilization will be determined by the time spent between the two arms. There is, however, nothing in the belt of life hypothesis to suggest that life will appear between the galactic arms; all that can rightly be inferred from the hypothesis is that any probability of intelligent life will be restricted to a smaller area of the galaxy than is predicted by the more optimistic estimates. One cannot argue from factors which would place a limit on the number of technological civilizations to a more accurate prediction of the number of such civilizations.

Further restrictions upon the probability of intelligent life have been introduced, by Antony A. Stark's (1988: 291–3) 'sleeping dragon' hypothesis, according to which the galactic centre is full of dense gases which become unstable every few million years or so. Intense bursts of Gamma-rays, X-rays and other energetic particles have been detected in the galactic centre. According to Stark, 'Every few million years the dragon awakes, and sterilizes the galactic centre with a burst of fire' (1988: 293). There is also a school of thought which supports the view that there is a gigantic black hole at the galactic centre. On these terms the development of civilizations will consequently be curtailed by periodic disasters, thus limiting the factor of L in regions located towards the centre of the galaxy.

According to James Annis (1999: 19), who is an astrophysicist at Fermilab near Chicago, periodic cataclysmic Gamma-ray bursts may sterilize entire galaxies, wiping out all life-forms before they have evolved sufficiently to either leave their planet or establish the ability to communicate with other planets. These Gamma-ray bursts (GRBs), he claims, are probably the most powerful explosions in the universe and are believed to be caused by collisions between either neutron stars or black holes. Annis maintains that each GRB unleashes massive radiation. If, for example, one went off near the centre of our galaxy, within seconds the Earth would be exposed to a massive wave of Gamma-rays that would exterminate almost every form of life. Even those which survived the initial zap of Gamma-rays, through shelter on the planet's dark side, would be wiped out by indirect effects such as the destruction of the ozone layers which protect from UV radiation. So it would seem that the effect of a GRB within a galaxy is that the evolution of life is pushed back to zero.

If this is the case, then the rate at which GRBs occur might profoundly influence the factors, f^i to f^c in Drake's equation, limiting the prospects of an emergent communicative network. The same limit would apply to developments in space travel and colonization, as a high incidence of GRBs throughout the universe would suggest that civilizations were unlikely to reach the threshold required for space travel. Apparently the rate of GRBs is one burst per galaxy every few hundred million years. Now this might allow a few civilizations to emerge and even serve as an incentive to communication and space colonization, as knowledge of future destructive episodes became apparent. However, Annis argues that the rate of GRBs was actually much higher in the past, with galaxies receiving one burst every few million years, which would almost certainly prohibit the evolution of an intergalactic civilization. Thus the reason we are not in contact with ETIs is that no civilization has been given enough time to develop that far.

This thesis has an intriguing implication: if GRBs are the main inhibiting factor on space communication and colonization, and so far they have not prevented the inhabitants of Earth from developing interstellar communication and reaching the threshold of space colonization, then other forms of intelligent life may be appearing in our galaxy and will continue to do so until we are all destroyed by the next GRB. So either we shall be in contact soon or we shall be annihilated when the sleeping dragon awakens.

Conclusion

The Drake equation is an edifice constructed from degrees of guesswork based on a range of theories which vary in evidential support. Only R* and f^p are supported by observations extending beyond our solar system. The equation makes no claims on behalf of the immunity of any of its component theories and inferences drawn from them carry only a theoretical possibility. Taken together each factor in the equation amounts to a chain of probabilities. Frank J. Tipler (1980: 273) has argued that 'the problem with the equation is that only f^p – and to a lesser extent n^c – is subject to experimental determination'. To measure probabilities with any degree of accuracy would require a large sample. But, argues Tipler, 'for f^c, f^i and f^c we have only one obvious case, the Earth' (ibid.). An unkind critic might say that the Drake equation is nothing more than a piece of inspired guesswork; comparable to guessing the number of stray dogs in Manchester by multiplying estimates of the canine birthrate by the number of food-containing waste-bins by a guess at the average life-span of a dog. In a more favourable light the equation might be said to provide a heuristic guide whereby new theories can be appended and various blanks filled in. Consequently the numbers which make up N will vary according to the theories currently used to underpin any of the stages in the equation and, of course, the personality of the scientist. Drake (1990), optimistically, argues that N = tens of thousands of communicating intelligences across the galaxy. Others are less optimistic.

At the other extreme, pessimists have assigned a value of 1 for N. The N = 1 civilization argument is based on a version of Fermi's Paradox (examined in Chapter 6), which maintains that if expansion into space is a consequence of advanced technology, then the probability is high that civilizations more advanced than ours have either visited the solar system or left a beacon nearby. Accordingly, the only way to account for the silence is to adopt a value of 1 for N.

Supporters of N = 1 also appeal to the element of chance in the stability of the Earth's climate, stressing its uniqueness as a condition for sustained life. But this view might be offset by appeal to the Gaia hypothesis (Lovelock, 1979) which argues that stability necessary for life is maintained in terms of open system interaction between the atmosphere, hydrosphere and the Earth's crust, and that certain feedback mechanisms – the evolution of microbes and marine organisms – contribute to the maintenance of a geothermal steady state. On such principles it is held to be probable that other biosystems are maintaining similar eco-systems.

There are various biologically based objections to the N = 1 school, which appeal to the ability of life to adapt to a wide range of habitats and claim that the emergence of creatures with mammalian intelligence is a repeatable phenomenon among numerous phyletic lines on Earth and is likely to emerge elsewhere.

What is the methodological status of the Drake equation? It is frequently said that it is more a way of organizing ignorance than a defining scientific methodology. According to Jakosky (1998: 285) Drake's equation is 'just a mathematical way of saying "who knows?" ' But he adds: 'It does allow us to focus on those issues that have the largest uncertainties: Does life occur on every planet where it is possible? Does intelligent life evolve as an imperative? How long can a civilization last?' It poses questions rather than answers; it does not provide proof, but it gives a structure to hypothesis generation which is based on falsifiable empirical developments in major branches of science including astrophysics and biology. Drake's structure conveys plausibility, which is not destroyed by the failure of any particular search; yet the component theories and empirical statements in Drake's structure are subject to either corroboration or falsification as knowledge develops. As long as this structure remains intact, it is possible to employ it when hypothesizing about intelligent civilizations. Yet it does not tell us where to look, and no extraterrestrial civilization can be deduced from it. It does not provide an indication of how many technological civilizations may exist; it is not an instrument that will tell us how to find them, but rather it brings together a series of problems to be solved by means of a fusion of many disciplines.

SETI, as Timothy Ferris (1992: 25) argues, is not so much a science as 'a campaign for exploration'. It has a broad emotional and metaphysical appeal, which accounts for its attractiveness, but it captures the excitement of the explorer rather than the investigative scientist. The difference between science and exploration is very fine, but basically science survives by making accurate predictions. Exploration does not: many of the great explorers have failed to provide even the remotest sign of an accurate prediction. The ancient Chinese navigated the Pacific in search of the elixir of life; Columbus predicted that he could sail west all the way to the Indies, which was impossible, especially as he believed that the Earth was merely one-third of its true size. We search for extraterrestrial intelligence, not because we have any hard knowledge of its existence, or know that we can find it, but because we think we might. Yet campaigns for exploration are an integral feature of frontier science, where the generation of untested hypotheses, new theories and experimental instruments are simultaneous with techniques of persuasion and appeals to future rewards.

The belief in the possibility of extraterrestrial intelligence as well as the Copernican principle, which displaces the idea of a unique Earth, functions like other metaphysical regulators in science. These beliefs, which are akin to what Sir Karl Popper describes as non-falsifiable conjectures about the world, have stimulated and directed scientific inquiry. They include some of the great scientific theories, such as the Cartesian theory of matter as a continuum, Democritus' atomic theory and Einstein's General Relativity Theory. What matters most is that the hypotheses derived from them are testable and conform to established scientific knowledge.

The authors of the Cyclops Report envisaged a search that would take centuries and spoke of the need for faith and perseverance as well as the latest technology.

> To undertake so enduring a program not only requires that the search be highly automated, it requires a long-term funding commitment. This in turn requires faith. Faith that the quest is worth the effort, faith that man will survive to reap the benefits of success, and faith that other races are, and have been equally curious and determined to expand their horizons. We are almost certainly the first intelligent species to undertake the search. The first races to do so undoubtedly followed their listening phase with long transmission epochs, and so have later races to enter the search. Their perseverance will be our greatest asset in our beginning listening phase.
>
> (Oliver and Billingham, 1973: 171)

Nevertheless, as a call for exploration the Drake equation appeals to many branches of science. $R*$ is the province of geophysics and astrophysics; f^p involves geophysics and atmospheric physics and n^e is to be examined in the boundary between astronomy and biology; f^l involves organic chemistry and biochemistry; f^i involves neurophysiology; f^c is bound up with anthropology, archaeology and history, while L is inescapably linked to politics, sociology and psychology. Although SETI is not taught in universities as a scientific subject and does not have its own scientific journal, it is very much in the field of the interdisciplinary empirical sciences bringing together work in astronomy, physics, engineering, biology and the social sciences.

3

HABITATION, LIFE AND
INTELLIGENCE

Space: the final frontier. These are the voyages of the Starship Enterprise. Her five-year mission: to explore strange new worlds, to seek out
new life and new civilizations, to boldly go where no man has gone before.

(*Star Trek*, beginning 8 September 1966)

Introduction

Theories regarding extraterrestrial intelligent life-forms must be supported by
evidence of potentially habitable sites, such as stable planets in a moderate
temperature zone. This chapter surveys contemporary claims concerning
potentially habitable planets, theories of planetary formation and methods of
observing ex-solar planets. The investigation then focuses on the development of
life, stressing an important distinction between conditions which may support the
origins of life and conditions which life can tolerate, reorganize or adapt. Given
the fact that there are probably many more sites where life can survive than
those where life can be produced from scratch, the thesis that life on Earth is the
result of seeding is defended as a more plausible account of the development of
life on Earth. Arguments concerning the contingency of life and the tenacity of
life are assessed and the emergence of intelligent life and exotic life-forms is also
considered.

The cosmological principle

Appeals to a high probability of intelligent life elsewhere in the galaxy provide
SETI research with its primary rationale. The appeal to large numbers, the
vastness of the universe, frequently referred to as the cosmological principle,
features strongly as a factor in SETI's plausibility. Ben Bova, a science fiction
writer, SETI enthusiast and former advisor to President Ronald Reagan on the
'Star Wars' project, expresses this belief: 'It is difficult to believe that around all
those billions of billions of stars we are the only living creatures, the only
intelligence' (Bova, 1990: 8). Jill Tarter taps the richness of the SETI programme

with an appeal to large numbers: 'we have explored so little of the Universe and the space around us that we cannot guarantee that "they" are not here' (1991: 185).

We have already argued that the appeal to the vastness of the universe is a good reason for optimistic expectations in favour of pluralistic hypotheses, and provides a good reason for conducting a search. But while such appeals may support a campaign for exploration, by themselves they carry no probative weight. The error in drawing definite conclusions from appeals to large numbers can be exposed with reference to our friends, the monkeys in the British Museum who have been set the task of randomly typing out the complete works of Shakespeare, given enough time and suitable word-processors. According to the mathematician, Frank Cousins (1972: 263), if we allow reasonable assumptions about typing speed the chance of an occasional Hamlet appearing is once in $10^{460,000}$ seconds, which is effectively never. Michael Crowe (1988: 553) illustrates the problem as follows:

> Consider a universe containing a billion galaxies, each made up of a billion stars, around each of which a hundred planets revolve. Place a billion monkeys on each of these planets and set them typing for fifteen billion years [the approximate age of the universe]. This would produce only 10^{46} seconds of typing time towards the $10^{460,000}$ seconds needed.

The intuitive strength of the appeal to an infinite amount of time and space lies in our familiarity with practical tasks. Richard Dawkins (1988: 139) appeals to the large number argument when formulating theories on the origin and evolution of life: 'Given infinite time or infinite opportunity,' he says, 'anything is possible.' But this claim is soon qualified with reference to what is, or is not, physically possible. Given enough time and a willing supply of driving instructors many an incompetent driver will pass the driving test. One might also predict that given time, the manufacturers of sun cream will provide a totally effective screen against UV radiation. But not even in an infinite amount of time will they provide a barrier that will enable people to take a holiday on the Sun. Appeals to infinity are obviously restricted, but they can indicate what is plausible within a particular theoretical domain. This point is made forcibly by Jean Heidmann who asks:

> How can we reject the idea that in billions of years, in billions of galaxies, each containing billions of stars, physical evolutionary processes may have produced more advanced results than those found in our small globe toward the end of what we insignificantly call our 20th century?
>
> (1997: 115)

It must be emphasized that this is an argument for the plausibility of a hypothesis; not a proof but an appeal for experiment. As Heidmann notes: 'SETI is, in fact, the only method at our disposal of trying to verify, by observation, that extraterrestrials exist' (ibid.: 116). The appeal to the vastness of the universe can be self-defeating. There are about 10^{10} galaxies in the universe. Suppose there are 10,000 million habitable planets, this would be roughly one planet for each galaxy. If this were the case, we would be unlikely to see them or even their parent stars. And if they were very far away, we could only see what they were several million years ago. Even to say that there are several thousand sites where ETI could flourish does not lend encouragement to expectations of an encounter. We need specific directions. If contact is made, it is likely to be due to luck.

Closely related to the appeal to large numbers is the appeal to analogies with the Earth. Given the vastness of the universe and a high probability of numerous Earth-like planets, then inferences drawn from Earth's natural history can be applied to theories concerning life on other planets. It is not clear as to how much can be inferred from Earth's natural history. Quite clearly the fact that Earth exists and supports intelligent life is proof that similar environments can exist, but in the absence of further evidence, claims concerning actual probabilities cannot be given much weight. This is not to say that such claims are valueless, or even unscientific, but that they belong to the generative stage of scientific inquiry rather than in the finished scientific report.

A frequent methodological problem in SETI research arises from the misuse of analogy, usually between the Earth and allegedly similar planets. Evidence for the existence of similar planets to Earth might be derived from appeals to the existence of similar G-type suns to our own. Without strong additional evidence such analogies have little probative value. This is not to say that analogy should have no role in scientific reasoning. In fact it has an important generative and heuristic role if properly applied (Lamb, 1991), but an appeal to analogy should not be offered as a conclusive proof. A total rejection of the use of analogical reasoning in science may, however, be a case of cutting off one's scientific nose to spite one's philosophical face. The use of what might be called 'timid' analogies is to be applauded. These only permit speculation around what is already scientifically possible. Thus, for example, speculation about possible extraterrestrial life might be contained within analogies drawn with sites where there is carbon-based life and liquid water.

A cautious use of analogical reasoning is clearly advised when appeals to the Earth's natural history are made in SETI research. Earth is not an adequate sample, as one cannot draw a statistical conclusion from a sample of one – although a sample of one is sufficient to refute a claim that something cannot happen. If extraterrestrial life exists, our knowledge of it cannot be based on inferences drawn from the Earth. While studies of the Earth can serve as a useful laboratory model for the development of life, they cannot provide reasons to suppose that other life-forms in the universe exist and will be intelligent. In short, there is no logical inference to extraterrestrial intelligence. Its discovery will have to be empirical, not by means of an appeal to logic and large numbers.

So much for the empirical requirements for SETI. But speculation and analogy need not be excluded altogether. One of NASA's ground rules for SETI is that life is likely to originate on planets orbiting a sun like ours. It is a fact that intelligent life evolved on one planet – the Earth – so the evolution of intelligent life cannot be ruled out *a priori*. After all, in 1491 most Europeans had no evidence that America existed.

If there is intelligent life elsewhere, the most sensible place to look for it is on ex-solar planets orbiting similar suns to our own. Life on other planets will require two necessary features. First, a stable source of energy from the parent star, a nearby gas giant or an internal heat source; second, plenty of raw materials in usable form. Although calculations abound concerning the possible number of planets like Earth which are capable of supporting life, until recently no planets have been directly observed beyond the solar system. It is widely believed that many stars have planets, but until recently ex-solar planets have proved extremely difficult to find, as they are likely to be distant and have no light of their own.

Ex-solar planets

The existence or non-existence of ex-solar planets is one of the most important issues facing SETI. Recent successes in the detection of ex-solar planets have led to optimistic forecasts. In 1999 Roger Angel at the University of Arizona forecast a 'revolution in planetary science' (Adler, 1999: 4). There is no doubt that many theories about the formation and composition of planets will be revised in the wake of new planetary observations. Nevertheless, evidence of ex-solar planets is not itself revolutionary, as most astronomers have believed in their existence for a considerable time. But without ex-solar planets there is little hope of making contact with other life-forms.

According to J.D. Scargle (1988: 79): 'The search for planets is the first step in the search for extraterrestrial intelligence, as the surface of a planet is probably the only viable location for the origin and evolution of life.' Since 1995 planetary masses have been detected by various means and we are now on the brink of discovering rocky Earth-type planets.

How common is our solar system? Is it typical of such systems throughout the universe? The answer to these questions depends very much on theories concerning the formation of stars and planets. Most astronomers believe that planets are commonly distributed throughout the galaxy, but until the end of the twentieth century there was no clear observational evidence beyond the solar system. Estimates of the number of possible planets have been based on a combination of intelligent guesswork, theories about the nature and formation of planets, and indirect observations of their alleged gravitational influence on other bodies.

Theories concerning the formation of stars and planets are called cosmogonic theories. The most acceptable theories at present are versions of nebula theory, according to which the star and its planetary system are formed by the same cloud of contracting dust. Since everything in the solar system rotates in the same direction it is plausible to assume that the entire system originated at the same time out of matter that accumulated to form the individual planets. On these terms both the Sun and its planets originated out of the same cloud of dust. But there are still many theories and cosmogony is highly dependent upon speculation, although progress would be made with developments in astronomical techniques which could facilitate more accurate observations of planets.

While current theory is suggestive of a very large number of planetary systems, accident or catastrophe theories of planetary formation, which were prevalent during the eighteenth and nineteenth centuries, placed restrictions on the number of possible planetary systems. In general such theories were governed by the belief that the solar system was unique. One version of the catastrophe theory was that the original binary companion of the Sun either destructed or collided with a third body, leaving debris which later settled into planets. The fault with this explanation lies in the fact that the result of the self-destruction of a star is usually a supernova with the debris hurled out too far to form orbiting bodies, and thus what remains is a white dwarf or neutron star. Collision with a third body would scatter matter in all directions, and would be unlikely to produce planetary orbits. Among the main exponents of accident theory was Compte de Buffon, who speculated in *De la formation des planètes*, in 1745, that the solar system was composed of matter torn from the Sun following an encounter with a comet. This speculation did not survive mathematical computation which reveals that even if a comet were large enough to cause great masses to leave the Sun, the gravitational separating force would be too great to allow condensation into a planetary system. The most likely result of such a collision would be an envelope of solar gases.

During the nineteenth century several astronomers were attracted to various 'tidal' theories of planetary formation. Thomas Crowther Chamberlain and Forest Ray Moulton developed a tidal theory which became known as the 'planetesimal hypothesis', according to which an alien star bypassed the Sun, drawing off matter which then condensed into planetesimals. By steady accretion these small bodies grew in size and then became planets. Objections to the hypothesis were based on doubts concerning the ability of planetesimals to endure the many collisions that were expected. It was more plausible to assume that they would be dissipated. Other problems with the hypothesis were a failure to account for planetary orbits, their distances from the Sun, and the rapid rotation of large planets like Jupiter and Saturn.

In the early twentieth century Sir James Jeans' and Harold Jeffreys' near collision hypothesis – a version of the tidal theory – was favoured. Jeans and Jeffreys maintained that a star passed near the Sun and pulled out a tidal filament from which the planets were formed. Given that the chance of stars colliding in our part of the galaxy is about one in a billion each year then the number of planetary systems was likely to be small. The main weakness of this hypothesis lay in explaining how the matter was to be drawn out of the stars as they passed each other. It is easy to imagine great tidal waves of matter drawn from each star as they attract each other, but the likelihood is that the matter would be drawn back to the original stars as they departed, not ejected freely into space. The thesis is now discredited because [1] the filament extracted from the Sun would not extend far enough; [2] the material drawn out would be too hot to condense into planets and would disperse instead; and [3] the very presence of deuterium and lithium on Earth suggests that planets were formed at lower temperatures than this model suggests.

Most current theories reject accidental explanations of planetary formation in favour of natural explanations which accordingly predict larger numbers of planetary systems. In some respects Kant and Laplace's respective nebula hypotheses have come back into favour, as both held that planets and stars originated together. In his *Allgemeine Naturgeschichte und Theorie des Himmels* (1775; 1981) Kant depicted the Sun as a huge slowly rotating body of gas filling up the entire volume of what was to become the solar system. As it cooled it contracted and spun more rapidly, reaching a point where rings of gases would break away and condense into planets. This hypothesis was developed by Pierre-Simon Marquis de Laplace, with an additional assumption of an extensive nebula surrounding the Sun. As the nebula cooled and contracted it rotated faster, such that centrifugal forces outmatched gravity and the rings fell off, later to condense, forming planets. The major objection was that the centrifugal forces would not be powerful enough to overcome the gravitational forces, and if they were, then the matter would be hurled too far away for it to coalesce into planets.

According to the currently prevailing, but not undisputed, theory, planets are natural by-products of the formation of a star from a collapsing cloud of gas and dust. This explanation is also supported by some of the more plausible accounts of the formation of an atmosphere according to which atmospheric constituents are already present in the dust grains. There is also evidence that organic material may be present in the stellar clouds. Planets are products of a flattened disk of material left over after the collapse of a slowly rotating interstellar cloud of gas and dust, consisting of silicon, oxygen and carbon. This would suggest that planets do not orbit the older – first generation – stars which consist mainly of hydrogen and helium and are not accompanied by clouds of dust. It is likely that ex-solar planetary systems will be found near the younger stars. Stars and planets are thus formed simultaneously as the result of the protostellar cloud collapse where the angular momentum of the cloud is small enough, but not exceeding the momentum required for the formation of stars and planets. The spinning clouds of dust can be triggered when they pass into a dense region near the leading edges of the galaxy's spiral arms, or by explosions of nearby supernovae (Hughes, 1992: 29). Eventually the cloud breaks up into smaller fragments. Determinant factors will be the rate of spin and the angle of spin. If it is spinning slowly, it may end up as a single star surrounded by low density gas and dust which will eventually disperse. But if it is spinning quickly, when it breaks up it will form a double or multiple star system. About half the stars have formed such systems. Somewhere between these extremes the condensing fragment becomes a star surrounded by a nebula which may become a planetary system. It is therefore important that stars, like our Sun, have peculiarly small angular momentum. Usually, massive stars have a high angular momentum while lower mass stars have less angular momentum and the formation of a planetary system may well be the means by which the loss of angular momentum occurs. According to one estimate about 'one in four and one in seven of all stars begin their lives with a nebula that has the potential to form planets' (ibid.: 30). There are clear limits on the formation of stars with planets and it has been estimated that planets form about 0.1 per cent of all stars of solar mass in the galaxy (Ruzmaikina, 1988: 41).

The actual mechanism by which planets evolve can be described as follows: as the dust cloud contracts, the heat generated by it is trapped and huge amounts of energy are released so it becomes a sun. Around this sun orbits a disk of dust and ice particles which collide with each other and eventually form together into small bodies called planetesimals. After several collisions they grow into planets in a rather wasteful hit-and-miss process. But this process accelerates nearer the star where mobility is faster. Thus in the solar system the Earth was formed in one million years compared to the about 300 million years which it took for the outer planets, Uranus and Neptune, to be formed. Data from the Hubble telescope lends confirmation to the hypothesis that the solar system is surrounded by a disk of icy comets, called the Kuiper belt (after the planetary astronomer, Gerard P. Kuiper, who proposed the existence of such a belt in 1951) of which Pluto and its moon, Charon, may be part. This disk may be the debris which never developed into planets. These observations suggest that tens of thousands of planetesimals may be orbiting the Sun (Horgan, 1994: 3–14).

Opinions vary concerning the nature of the clouds of dust which have been observed around certain stars. Are they the building blocks of planets or are they, as some astronomers suggest, what remains after the formation of planets is completed? Disks have been observed around young stars which supports the suggestion that the disks are themselves part of the early process of planet formation. The Hubble telescope revealed a number of young stars surrounded by disks of dust which could be indicative of the early stages of planetary formation. Further supporting evidence of the nebula hypothesis has come from infrared observations of a young star, MWC 349, which some astronomers believe to be an example of the primordial nebula that produced the solar system. This particular star has about 30 times the mass of the solar system. Its luminosity appears to have decreased by 88 per cent during the forty years it has been observed. This is now attributed to the dissipation of disk material (Brahic,

1994). The most striking claims so far were made during the mid-1980s when astronomers observed a disk around the young star, Beta Pictoris, and similar reported observations were made regarding the T Tauri stars in the constellation of Taurus. It is believed that T Tauri stars have disks of dust around them which might be in the process of generating a planetary system. Evidence for this was derived from infrared radiation which is believed to come from the dust particles in the disks. Older stars do not appear to show signs of infrared radiation, so it is believed that the dust has settled into planetary masses. On the other hand, the dust might have been blown away by the stellar winds, i.e. streams of subatomic particles ejected from stars at high velocities. First-generation stars, consisting of hydrogen and helium, would be unlikely to be accompanied by a dust cloud.

A planetary system would normally be restricted to single stars, neither too large nor too small, with medium angular momentum. Although there still remains great uncertainty with regard to the formation of multiple star systems, their complicated gravitational systems indicate that they are unlikely places to find planetary systems – although they should not be ruled out altogether. About half the stars in the universe are in multiple systems. David Hughes (1992: 33) calculates that about one in twenty-four stars could have planetary systems and that the average number of planets around stars that could have them is 13.5. This is a possibility of 600,000 million planets in our galaxy which contains about one million stars.

It should be stressed that during the early 1990s such calculations and theories provided hints not evidence regarding ex-solar planets. These hints, however, had an accumulative effect. The Hubble telescope pictured several disks in the Orion Nebula. Supportive evidence also came from the IRAS (Infrared Astronomical Satellite) which found evidence for the proto-planetary nebulae around several nearby stars, such as Beta Pictoris. It would appear to be a very young disk and it is not known whether planets have evolved there yet, although spectroscope observations indicated that bodies of a few kilometres in diameter may be present in the disk (Vidal-Madjar, 1994: 415). Further hints were to come from observations of the rotation rates for stars. Our Sun, like other G-type stars, rotates slower than some of the large stars. One explanation is that it has been slowed down by the gravitational effects of the planets. Very large fast-rotating stars will have drawn their planets into them, like a skater who draws her arms inwards and spins faster. This feature increases the plausibility of the hypothesis that slowly rotating stars have planets. But critics might argue that all of this is based on an unfounded assumption of a uniform rate of rotation for all stars. In fact, there is evidence that some T Tauri stars with a similar mass to our Sun rotate as fast as the larger stars. Moreover, there are alternative explanations for the slow rotation rates of certain stars. Stars, like our Sun, for example, may decelerate their rate of spin through the gradual interaction of stellar winds with their magnetic fields, with the winds acting as brakes.

Observing ex-solar planets

One thing that emerges from the variety of theories about the relationships between stars and planets is that the existence of ex-solar planets cannot be deduced simply from theories. Confirmation must be derived from observation. Observations of ex-solar planets can be categorized in terms of direct and indirect strategies. Direct strategies involve trying to extract a faint planetary image from the bright image of the star. The advantage of this method is fast verification with no requirement to observe a complete orbit, which could take hundreds of years. However, direct observation is extremely difficult. The Earth would not be discernible with a telescope even one-hundredth of a light year away. Planets do not emit light and their size in relation to their parent stars will be very small. Jupiter, the largest planet in the solar system, is 300 times the mass of the Earth but 1,000 times less than the Sun. Light from the Sun is about a billion times brighter than the light reflected from Jupiter. Using the best observational equipment available, an astronomer on one of the nearest stars – for example, Proxima Centauri which is 4.2 light years away – would not be able to distinguish Jupiter from the Sun. It would be lost in a glare of sunlight.

There are several promising techniques underway. For example, techniques known as 'optical' and 'nulling interferometry' aim at suppressing the glare of starlight that would otherwise conceal planets. This involves the use of two or more telescopes located at some distance from each other thereby gaining the resolution of a single telescope. Optical interferometry, although very complex, will detect a star's wobble caused by the surrounding planets and even directly image these planets and possibly analyse their atmospheres. Nulling interferometry will employ two mirrors to collect light from a star and combine it so that the light from one mirror is slightly out of step with the light from the other. The idea is that the light waves from the two mirrors slightly suppress each other, while the light coming from objects near the star is not delayed. Plans are underway for the launch of the Space Interferometer Mission in 2005, which will employ six telescopes aligned with mirrors that ought to detect or rule out the existence of Earth-type planets around the nearest twenty or so stars.

Can planets be observed indirectly? Indirect strategies involve methods of sensing the effect the planet may have on the star. Two kinds of indirect approaches were outlined by J.D. Scargle (1988). First, there was the Photometric Planetary Detection Method. If the planet transits in front of the star as seen from Earth, the light from the star will be diminished. This would create problems for detection as one would have to view the orbital plane edge on. Consequently this technique is only applicable to planets whose orbital plane is near our line of sight at the time of transit. One major problem is that this technique is non-repeatable, as the star, the intervening object and the Earth, are unlikely to be aligned again. Furthermore, there are problems involved with attempts to separate luminosity fluctuations from the star itself from those caused by the planet. This would be particularly difficult in the case of small planets like Earth, although large planets like Jupiter can be detected with this method.

Since 1994 a binary star system, CM Draconis, which is about 60 light years away, has been observed by a research team who have employed ten telescopes distributed around the Earth. Several astronomers recorded a dimming of the star's light during what might be inferred as transits as a small planet passed in front. It now seems that the scientific community are prepared to accept that the observational technique actually works. On these terms proof of the existence of ex-solar planets is not via direct observation, but follows predictable observations, based on calculations of planetary orbits determined by regular dimming of the light from the parent star (Seife, 1998a: 23). A most promising proposal is the photometric space-based telescope known as KEPLER, which is planned for 2003 and will orbit the Sun seeking potential alignments between ex-solar planets and their stars. It is hoped that KEPLER will find 500 rocky Earth-type planets, 160 close-orbiting giants and 24 outer-orbiting gas giants, during its proposed four-year programme.

The second indirect method is based on measuring gravitational perturbation. This is known as astrometry – the detection of the sources of gravitational pull on stars. Planets can influence their parent stars by means of their mass. Jupiter certainly affects the Sun's motion. Some nearby stars appear to wobble as they move through space which suggests that they could be gravitationally affected by unseen planets. Barnard's star, which is the second nearest star to the Earth, was widely proclaimed to have one or more planets when Peter van de Kamp, an astronomer at Sproul Observatory, announced the discovery of a wobble in 1963. It was then inferred that it had a 'companion' which was about 50 per cent heavier than Jupiter. Unfortunately, a fresh look at the original measurements suggested that the 'wobble' was due to a series of mechanical adjustments which had been made to the telescope.

Many searches are now underway, and over thirty ex-solar planets have been detected since 1995. There are also estimated to be numerous 'dark companions' among the nearest forty or so stars. However, there are difficulties in measuring the wobble. Although measuring of gravitational effects is one of the better indirect methods, there is a major problem in separating the effects caused by the planet from other motions in the stellar atmosphere. Moreover, this might only be possible in the case of very large planets. For example, the star's position would have to be measured relative to a set of different stars seen in the same direction. But these measurements are unlikely to be accurate enough to detect the influence of a planet. It would be impossible to measure the influence of Jupiter on the Sun from an observatory in Proxima Centauri. It should also be noted that full confirmation of a planetary orbit can take a long time. For example, the planet, Jupiter, takes twelve years to orbit the Sun, and astronomers beyond our solar system would have to spend decades confirming its existence.

At the University of British Columbia Bruce Campbell, Gordon Walker and Stephenson Yang conducted a twelve-year-long search using the 3.6 metre Canada–France–Hawaii telescope on Mount Kea, Hawaii. They employed a 'Doppler technique' which involves observing the motion of a star and measuring the frequencies of its light to determine how fast it is approaching or receding (Campbell, 1990). The 'Doppler effect' is the observed change in the wavelength emitted by a star due to its motion. Light from far-off stars, which are moving away from us because of the expansion of the universe, has a longer wavelength. This is known as the 'red shift'. If the star were moving towards us it would exhibit a shorter wavelength or 'blue shift'. A tiny fluctuation from a red to a blue shift indicates that something is causing the star to wobble.

The distribution of angular momentum (or the amount of spin) could provide indirect evidence of a planetary system. A slowly spinning star could be so because of the gravitational effect of its planets. The rate of spin can be detected by a study of the light it emits. Its movements affect the frequencies of light that reaches us due to the Doppler effect. A star with an orbiting companion will oscillate as it goes into its counter-orbit. This can be seen in the case of double stars. But it is harder to detect any change in velocity with presumed planets, as the oscillatory effect would be smaller. However, during the 1990s improvements in observation techniques, including spectral analysis, enabled fairly accurate measurements of the velocities of stars and their speed changes as low as 30 mph. Doppler techniques were to prove successful, leading to a spate of observations at the close of the twentieth century. Many of these announcements followed a search for wobble which showed up as Doppler shifts in a star's spectrum due to the gravitational influence of orbiting planets.

Until 1995 no true observations of ex-solar planets had been detected, but by the end of 1996 more than twelve had been confirmed and more were expected. The American, David Latham, of the Harvard-Smithsonian Center for Astrophysics, with Michel Mayor at the Geneva Observatory, discovered that the star, HD116762, which is about 90 light years away, has a regular oscillation indicative of a massive planet about eleven times the size of Jupiter which completes its orbit in eighty-four days (Heidmann, 1995: 17). In October 1995, Mayor and Didier Queloz of the Geneva Observatory detected a wobble in the star 51 Pegasi, which is held to be caused by a Jupiter-sized planet (Mayor and Queloz, 1995). In January 1996, Geoff Marcy of San Francisco State University and Paul Butler of the University of California claimed to have detected wobbles caused by two planets; one in the constellation of Virgo which is about 6.5 times the mass of Jupiter and another in the Great Bear, which is allegedly 2.3 times the mass of Jupiter. Both are about 35 light years from the Earth. In January of 1996 Chris Burrows of the Space Telescope Institute in Baltimore, Maryland, cited supporting evidence of a planet from an image taken by the Hubble telescope which revealed a slight warp in the disk around Beta-Pictoris, which is about 50 light years away (Walker, 1996).

In June 1998, scientists reported a planet about 1.6 times as big as Jupiter orbiting the star, Gliese 876, which is 15 light years away. The alleged planet was detected by observing the 'wobble' of its parent star. The Gliese is a very low mass star only about one-third that of the Sun. But if there are planets so close to the solar system near an uncharacteristic star like Gliese 876, then this would suggest that planetary systems are fairly widespread. So far, it has been mainly gas giants that have been observed, but large rocky planets are being discovered as techniques improve, and the detection of Earth-type rocky planets cannot be far off. Early in 2000 astronomers using the Keck Telescope in Hawaii found one rocky planet less than the size of Saturn orbiting the star HD46375, which is 109 light years away in the constellation Monoceros, and another planet of similar size around 79 Ceti, which is 117 light years away in the constellation Cetus. These discoveries suggest that the existence of smaller rocky planets may be fairly widespread.

An analysis of a star's Doppler shift in December 1999 preceded claims that a planet estimated at twice the diameter and eight times the mass of Jupiter orbits the star, Tau Boötis, which is a Sun-sized star 51 light years from Earth. Its orbit is so close to the parent star that its atmosphere is 1500°C, which, according to prevailing theory, is hot enough for its clouds to include molten iron, as well as sodium and potassium (Adler, 1999: 4). This means that the planet will absorb most of the light reaching its atmosphere. A team of astronomers from the University of St Andrews employed spectral analysis to separate the light emanating from the planet from the glare of Tau Boötis, using data from the 4.2 metre Herschel telescope at La Palma (ibid.: 4).

The detection of planets is of clear benefit to SETI research, and will have a profound effect on current theories of planetary formation. Unfortunately, if a planet is large enough to affect the motions of a star to any significant amount, it would be an unlikely candidate for life similar to that found on Earth. Observational methods leading to the satisfactory detection of Earth-sized rocky planets will be greatly enhanced by successors to the Hubble space telescope and space platforms, although major technical problems still have to be solved with regard to the need to separate images of the planet from the background light of the star. For this reason, ground-based telescopes are limited to observations of wobbles and shadows of planets larger than Jupiter. However, an array of infrared telescopes on a space station could detect Earth-sized planets directly, which are currently obscured from ground-based telescopes by the glare of scattered light from their parent stars. Proposals are underway for a 50-metre array of four infrared telescopes which could be situated near Jupiter's orbit, at an estimated cost of US$2 billion (Hecht, 1996).

Even with advances in optical equipment there are still major problems in the identification, eradication and control of systematic errors. But dramatic future developments in the technology required for planetary observation should not be ruled out in the immediate future. During the 1840s Auguste Comte outlined limits on our knowledge of heavenly bodies, stating that we could never know anything of their chemical or mineralogical structure. This was but a few decades before the invention of the spectroscope. Nowadays analysis of the chemical and mineralogical features of objects in space is one of the central tasks conducted by astronomers.

Claims have been made on behalf of techniques known as 'adaptive optics' as a means of detecting planets (Wilson, 1995). This technique was developed by the US military as a means of spotting Soviet satellites in space. Adaptive optics involves techniques designed to correct the distortions created by the Earth's atmosphere. The objective is to create a mask to absorb most of the light from the central star. This involves the employment of four large deformable mirrors at least six metres across which can suppress the halo around a star's image so that a telescope could produce a clear crisp picture of the observed object and could pick out giant planets after only a few hours of watching (J. Davies, 1995: 24). Although this technology is in its infancy, future developments suggest that the increased sensitivity it offers to ground-based observations may make it possible to observe planets the size of Jupiter around stars in the Milky Way. One discovery relating to this technique is a companion to the star, Gliese 229.

At present it is difficult to maintain a distinction between planets and brown dwarfs. This is not merely a problem relating to observational techniques; it is bound up with theories concerning the origin and nature of both planets and brown dwarfs. Brown dwarfs are not as bright and as large as stars, and very often cannot be distinguished from gas giants. It is not clear whether they undergo the same formation process as giant planets, and they may arrive at similar ends via different processes. But if many of the alleged planets turn out to be brown dwarfs, then expectations of large numbers of ex-solar planets will have to be reduced. It is also likely that as more planets are detected with counter-claims that they are brown dwarfs, then the pressure to distinguish between them will itself generate both theories and observations which will reveal diversity with regard to their formation. It is safe to predict that several theories regarding the origins of planetary systems will be called into question during the next few years.

Already it is becoming apparent that very few of the observed ex-solar planets exhibit features resembling our solar system. Early in 2000 astronomers at the University of Oxford and the University of Hertfordshire in England discovered thirteen planets with masses less than Jupiter. What was significant about these planets is that they are free floating; they drift through space by themselves and do not orbit any star. Another feature of recently observed planets is that many of the giant planets are much closer to their stars than Mercury is to the Sun. Yet according to the prevailing model of planetary formation, small rocky planets, like Mercury, Venus, Earth and Mars, should occupy the hot inner regions, while the gas giants in the outer regions are able to grow large by virtue of their longer orbits which enable them to collect more material. The fact that so many of the gas giants recently observed have extremely close orbits around their stars may require further explanation. It appears that the detection of ex-solar planetary systems presents a significant challenge to the standard model of planet formation.

Among the searches for ex-solar planets is NASA's Astronomical Studies of Extrasolar Planetary Systems (ASEPS), while the European Space Agency (ESA) has identified planet detection as a priority and is considering several rival planetary detection methods. A proposal called DARWIN, by ESA in 1993, was for a search of signs of planetary life using a space-based interferometer consisting of several infrared telescopes which could detect a planet with oxygen in its atmosphere. Its rival was a programme called GAIA using astrometric techniques. There is also a proposed search for the Frequency of Earth Sized Planets (FRESIP) by the NASA Ames Center in California. The objective of FRESIP is to monitor the brightness of stars to search for changes caused by planets crossing them. This may be a small effect, but it can last for several hours and happens once a year. Such a regular feature would rule out random factors and allow predictions to be made regarding the planet's transit. The programme requires round-the-clock monitoring and will have to be conducted from a space station, where there is no interference from daylight or bad weather. Bill Borucki of NASA's FRESIP proposes a search of 5,000 Sun-like stars. If planetary systems are common, FRESIP should detect about fifty transits each year (J. Davies, 1995). Certainty in the search for ex-solar planets will be enhanced by means of the Terrestrial Planet Finder, due to be launched in 2011, which will combine light from four giant telescopes to create sharp images of planetary systems.

All of the above-mentioned developments in the methods of detecting and observing ex-solar planets, and their prospect of success indicate a shift from merely listening for ET signals to an active search for potentially habitable sites. Although the detection of potentially habitable planets is not in itself supportive of ETI hypotheses, common sense suggests that these sites are the most promising places to concentrate future searches.

Pulsars and planets

Several claims have been made regarding the detection of planets which, so it is claimed, orbit pulsars. Pulsars are the result of the explosion of massive stars. When the core of a star collapses, the atomic particles may be squeezed so hard that they coalesce to form neutrons. A star composed almost entirely of neutrons is called a neutron star. Many of these are pulsars, emitting powerful beams of radio energy from their poles in regular time bursts as they complete a full rotation in a fraction of a second. They offer a unique opportunity to determine if planets orbit them, as they emit energy in regular pulsating bursts. Each radio pulse is 10 to 20 milliseconds long every 1.33730113 seconds. Hence an interruption by a regular orbiting body can be easily detected. Thus when the beam passes through the Earth it is detected as a radio pulse. As these pulses are extremely regular, any variation in the timing of pulses can be accurately detected. Reflex motion, caused by gravitational attraction of an orbiting mass, will affect the timing of the pulse.

In 1991 Aleksander Wolszczan, of Arecibo Radio Observatory in Puerto Rico, and Dail A. Frail of the National Radio Observatory in New Mexico, reported two planets around the pulsar PSR 1257 + 12, which is about 1,600 light years away (Wolszczan and Frail, 1992). They were estimated to be about three times the size of the Earth. Pulsar PSR 1257 + 12 was observed to speed up and slow down over a period of 25.3 days, and Wolszczan suggested that this was the effect of an orbiting planet. He also claimed that two more distant planets were affecting the pulsar's rotation every 66 and 98 days. Three years later sceptics put forward an alternative view, that the change in the signal from the pulsar was not due to the effect of a planetary mass, but rather was caused by a coronal hole in the Sun out of which comes a fast solar wind which periodically slows down radio pulses as they travel through space. This view has been contested and the matter has not been finally resolved (Seife, 1997: 12).

In the summer of 1991 scientists at Jodrell Bank, using the same pulsar timing method as the Arecibo group, claimed to have discovered a planet about ten times the size of the Earth orbiting a pulsar known as PSR 1829 – 10. However, in January 1992, Professor Andrew Lynn of the Department of Radio Astronomy, University of Manchester, acknowledged that they had been misled by an error in their calculations. Apparently they forgot to take account of the eccentric character of the Earth's own orbit (reported in *The Guardian*, Katz, 1992).

The status of pulsar orbiting masses is unclear, and at the time they were first observed there were doubts among astronomers whether these alleged masses were actually similar to planets in our solar system. Ivan King, Professor of Astronomy at the University of California, Berkeley, argued that planets which orbit pulsars will differ greatly from planets like Earth: 'They do not have anything to do with ordinary planets going around our Sun ... the question of whether planets exist around ordinary stars is a completely unsettled one' (quoted by Crosswell, 1991b: 11). In any case, there is no possibility of life on the 'planets' around PSR 1257 + 12, as it gives off deadly radiation. Moreover, the heat from it is baking the two inner planets to a temperature of many hundreds of degrees Celsius (Crosswell, 1992b). While pulsars are not inimical to life, it was nevertheless argued that if planets can be formed around pulsars they can also form around stars. Thus the detection of planets orbiting pulsars supports theories of accumulative processes in circumstellar disks. The observation of pulsars therefore provides the best confirmation of current theories of planetary formation which in turn favour the more generous predictions regarding ex-solar planets.

There are suggestions that pulsars would be promising targets for intelligent radio signals. According to Jean Heidmann (1997: 168–70), because pulsars are natural interstellar radio beacons, powerful and well distributed throughout interstellar space, and emit pulses of staggering regularity, and have a lifetime of a million years, they would make excellent cosmic lighthouses for possible interstellar civilizations. Heidmann suggests that we listen at the frequency of a selected pulsar multiplied by a universal mathematical constant which will obtain a frequency at the 1 to 10GHz SETI window. Heidmann has actually submitted this proposal to an observational test and achieved an alert in the form of a strong signal from the region of the star, DM − 2398646. Unfortunately, when monitored three days later no repeat was detected.

Optimistic expectations regarding the numbers of ex-solar planets have encouraged SETI researchers. Steven J. Dick has concluded that the search for planets this century has consolidated a cosmological world-view that 'assumed that planetary systems were common, that life has developed on many of those planets, and that intelligence may have evolved to the point where we can communicate by radio waves' (Dick, 1996: 221).

Life

What is life? There are few definitions which are free from both scientific and philosophical contestability. The concept of life is hard to define, and no satisfactory definition has been proposed that has sufficient universality to cover extraterrestrial life. However, a useful pragmatic definition might be 'any object which feeds and reproduces'. These functions are so widely associated with life that their existence provides a conceptual foothold for considering something that feeds and reproduces as a candidate for life. One might find that this includes self-fuelling and replicating robots, and this might cause philosophical problems, but they could be regarded as examples of life of some sort, no more unusual or marvellous than some of the millions of living terrestrial objects. Of course the origin of robotic life – designed by some form of intelligence – can be apparent, whereas the origin of natural life is still unresolved. One approach to understanding the nature and origin of life is to consider the conditions favourable to its emergence and survival.

If there are many ex-solar planets, then what are the chances of life on them? The minimum condition for life of the terrestrial chemical type is a thermal range bounded by the liquidity of water, on the one hand, and the instability of proteins and other chemicals at high temperatures, on the other. There is, however, an important distinction between conditions which are favourable to the emergence of life and conditions to which life can tolerate or adapt. Many living beings can learn to cope with temperatures slightly below -20°C and over 100°C, but conditions suitable for the generation of life would be found somewhere in the middle of this range. Thus Mars, with extremes of low temperature, might not be capable of generating life, but under appropriate protective conditions it could sustain life. Bacteria and plants have been known to survive in simulated Martian conditions. Cacti have been grown in an assimilated Martian environment. Two other plants exposed to a simulated Martian environment for one month at the United Carbide Research Institute, New York, died due to temperature extremes but vigorous young shoots flourished from the dying mature plants (McGowan and Ordway, 1966: 83). It may be objected that these simulations are largely based on speculation regarding the nature of the Martian environment. Nevertheless, they do indicate that life is more adaptable than it was formerly thought. If manned space exploration continues, it may only be one or two generations before (human) life exists on Mars.

The best conditions for the development of life are limited to planets located in a habitable zone around each star. Earth is the prime example. Applying the Goldilocks criterion to our own solar system, Venus is too hot and Mars is too cold. Therefore, life-sustaining planets would have to exist within a narrow thermal range. Life on Earth is precarious. A slight fluctuation in the Sun's energy would wipe out most forms of life. A little hotter and the oceans would dry up; slightly cooler and the planet would be icebound. Fortunately the Sun has been stable for 5 billion years and it is likely to remain so for another 5 billion years. This is not the case with many other stars, as only the F, K, and G types are stable. Our Sun is a G-type, but the massive O-type stars, which burn fast and only exist for 10 million years as stable power sources, are not likely to be around long enough to support the evolution of life – certainly not in any advanced form. The size of the star is therefore extremely important. Generally the larger the star, the shorter is its life. The evolution of complex life-forms on Earth took 2 billion years. A star 50 per cent more massive than our Sun will only last 2 billion years before swelling up to become a red giant engulfing its planetary system.

Not only does life depend on the stability of the Sun's energy output, it also depends on the fact that the Earth's orbit is relatively circular, which guarantees stability of temperature. Thus a minimum condition for sustained life is that a planet remains in a thermally habitable zone. The larger the zone the better.

There is growing evidence regarding the existence of planets which orbit binary star systems. Astronomers have found evidence of a planet orbiting a binary star system in Sagittarius, which is 20,000 light years away. As between one-half and two-thirds of the stars in our neighbourhood are binary, this would suggest that planets are very widespread. However, planets associated with double star systems are generally unsatisfactory candidates for life, as their complex orbits are likely to take them out of a thermally stable zone. Irregular orbits would not have the strict constancy of solar heat, and this would rule out most stellar binary systems, which are the majority of stars. Nevertheless, there are some chances of stability, as some binary stars have close circular orbits. If close together, their planets could have a stable orbit as the stars' gravitational effect would be approximate to each other. But if apart, then the planetary orbit would be such that they would at times be closer to one star than the other, not only experiencing varying rates of energy from the stars but adding to the risk of planetary collision. Although life could possibly survive in such a situation, extremes of temperature and risk of collision make it less likely than in a stable single star system. However, if we eliminate all double and multiple star systems from our calculations, this could still leave about 20 billion stars in the galaxy which are stable and could support life if they had planetary systems.

The size of the planet is also important. If too small the planet would not have enough mass to retain seas and consequently there would be no atmosphere; as in the case of our Moon, where there is no liquid, and any gas emanating from the rocks will evaporate. If the planet is too large, like Jupiter, then the sheer quantity of atmosphere, its density, would prevent sunlight reaching the surface which would have an adverse affect on evolution. In a dense atmosphere it would take longer for the greater quantity of methane and ammonia to be replaced by oxygen, which would also inhibit evolution. However, a very large planet like Jupiter might provide its own internal heat in sufficient quantity to sustain the development of life on one of its satellites. It might be speculated that the newly discovered giant planets have satellites which could be almost as large and as rocky as the Earth. Whether or not one of the gas giants can capture an Earth-sized rocky planet is likely to remain a matter of conjecture for some time.

The planet's rotation may be a crucial factor: if the speed of rotation is too slow it would heat up during the day and freeze at night, like Mercury, which turns on its axis once every two months. Thus in the day its temperature can be as high as 430°C, and at night it is -170°C, which is hardly compatible with the emergence of life. If the rotation is too fast it would create a strong atmosphere, even shedding matter into space. Even the inclination of the planet's axis will determine how much of its surface is habitable.

Other factors may interrupt the development of life or extinguish it alto-gether. Dramatic solar flares, small variations in solar energy or supernovae occurring close to the planet so as to smother it with high energy radiation would inhibit the development of complex living systems. This has already happened more than forty times in the Earth's existence and it is likely that the develop-ment of life on Earth has been interrupted many more times. Comets or large meteor impacts might wipe out most living beings, but even small impacts may cause dust clouds, poison water, and change the environment in ways that prohibit or interrupt the development of life.

Among the claims for ET life the most pervasive methodological fallacy is mistaking necessary conditions for life for the sufficient conditions. A suitable temperature and air are two necessary conditions but they are only two out of numerous other conditions, all of which are required for an environment to be habitable. Yet all too often, evidence of oxygen in an appropriate thermal range has been proposed as evidence of ET life. But such evidence is only conclusive if combined with other facts that together constitute the sufficient conditions of life. The materials necessary for life exist throughout the galaxy. Given a stable environment, the building blocks of life, such as the amino acids and other precursors of living matter could be formed out of methane, ammonia and cyanogen, which astronomers have detected in large clouds of complex molecules in space. The problem is whether or not all the other conditions which are sufficient for the development of life are present.

It is not known if there are any other planets that could sustain life, but even if they could we could not infer that they actually do sustain life. Percival Lowell's (1909) belief that Mars was inhabited rested partly on empirical claims regarding a complex system of canals but mainly upon an assumption that life and the conditions for supporting life always coincided. From observations that living phenomena and the environment change together he inferred that 'Life and its domicile change together', adding that 'life but waits in the wings of existence for its cue, to enter the scene the moment the stage is set' (Lowell, 1909: 69). From the assumption that life will accompany the conditions which support life, Lowell was able to secure his hypothesis that Mars was inhabited by an appeal to empirical claims regarding a favourable atmosphere, signs of vegetation and so on. This inference is fallacious. We cannot infer actual life from evidence that a planet is habitable. I *could* live in Spain but I do not. Nevertheless, this view concerning the necessity, or predestination, of life has been echoed by exobiologists throughout the twentieth century. Thus it is argued that if conditions favour life it will emerge, and if it emerges then it will lead to intelligence. In reply it must be asserted that proof of the existence of extraterrestrial intelligence must be direct; it cannot be inferred from descriptions of conditions in which intelligent life would be comfortable.

One of the problems in identifying life is that early forms of life do not have the qualities associated with more advanced forms. No definition can provide a fine demarcation line between life and non-life at the primitive molecular level. Yet one of the preconditions of advanced forms of life is the emergence of multicellular organisms. For most of the Earth's history life was unicellular. Multicellular organisms appeared about 700 million years ago.

One thing is certain; life has evolved on Earth and continues to do so. Moreover, life on Earth can survive in extremely inhospitable conditions, such as the bottom of the oceans, in frozen wasteland and on mountain tops. In this respect Earth can be viewed as a laboratory which, if the laws of physics and chemistry are universal, can provide experimental models of how life can develop. But Earth is not the only laboratory. Suggestions have been made that Saturn's satellite, Titan, could be investigated as a precursor site for emergent life. When the Voyager spacecraft, launched in 1977, flew past Titan, it revealed spectrographs showing the presence of organic molecules, including complex hydrocarbons, acetylene, and hydrocyanic acid, which are fundamental building blocks of life. Titan's diameter is 30,000 miles, half that of the Earth; it has a dense atmosphere, essentially composed of nitrogen. This would be the case on Earth but for the presence of oxygen of biological origin, which is mixed with the nitrogen. But because Titan is far from the Sun its temperature is $-170^{\circ}C$. Titan could be a good model of a primitive Earth in a state of deep freeze (Heidmann, 1989: 59).

The origins of life

Our galaxy is at least 12 billion years old. The Earth is 4.6 billion years old. Life on Earth can be traced back to about 3.85 billion years ago. Rocks 3.5 billion years old contain fossil organic molecules suggesting a rapid emergence of biotic activity. There are stromatolites, found at the North Pole, about 3.8 billion years old, and perhaps even older ones have been discovered in Western Greenland. Life on Earth must have appeared fairly quickly, less than 700 million years after the planet's formation. The early atmosphere contained an abundance of biogenic compounds but very little oxygen. There are fossil remains of the earliest oxygen-breathing eukaryotic cells which have been found in Siberian rocks about 725 million years old and about 800 million years old in the Grand Canyon (Gribbin, 1994: 92).

How life emerged is very much a matter of speculation. There are three possible options: [1] it began as a supernatural act; [2] it emerged from natural physico-chemical processes; or [3] it came to Earth from external sources. The first suggests an arbitrary change which, like accounts of miracles, is not refutable, but is not compatible with scientific inquiry. The second suggestion can be pursued, using available knowledge about the development of life on Earth. At a chemical level life involves the interaction between nucleic acids and proteins. Darwinian theory stresses that the root of all life is the 'primordial germ', from which more complex forms developed. In the 1930s John Haldane in England and Alexander Oparin in the USSR suggested that life was originally formed in warm volcanic oceans in the early Earth. They spoke of a 'warm dilute soup', but our knowledge of this phase is very hazy as there are no traces left of the period. There has been no success in showing that these alleged early conditions actually occurred. Biologically significant small molecules have been formed in experiments, in the 1950s and 1960s, under conditions believed similar to the primitive Earth.

Some SETI exponents often appeal to the following argument. Life began with a combination of certain chemicals with carbon and water, which are abundant throughout the universe. If life appeared within the first billion of Earth's 4.6 billion-year history, it seems likely that it will occur on similar planets, given a favourable environment. Support is also drawn from laboratory experiments in which the early conditions on Earth are replicated. For example, in 1952 Stanley Miller and Harold Urey successfully synthesized amino acids in large quantities and thus demonstrated that a primitive atmosphere of methane, ammonia, water vapour, and molecular hydrogen bathed in ultra-violet light and charged with electric shocks akin to lightning, can replicate conditions whereby the early building blocks, 'precursor' molecules, like glycine and alanine, are formed. Various formulations of the Miller–Urey experiment have succeeded in synthesizing 18 of the 20 amino acids needed for life. Yet despite some initial successes involving the synthesis of polymers and more sophisticated precursors, attempts to simulate the 'accidental' conditions in which life allegedly started have not gone far. Critics of the 'warm ponds theory' frequently point out that scientists have so far failed to duplicate the spontaneous origins of life in laboratory conditions. However, the theory might be defended by using assumptions familiar to SETI. Life has had millions of years to evolve spontaneously on Earth, or even in some other region. But attempts to duplicate it in the laboratory have only involved a few thousand 'man-hours' to date. So put a few thousand million chemists in a few thousand million laboratories for a few million years and, so it is argued, life will be duplicated in a test tube.

Here, however, the appeal to large numbers lacks credibility because no effective model or theoretical framework is provided to demonstrate the difference between insurmountable problems and practical problems. Critics have pointed out that there are no satisfactory models of the origin of life. It is no longer believed that life occurred out of a jumble of chemicals in a soup plus energy, and it is known that the characteristics of life are found in the logical structure and organization of the complex molecules. But to date we have no idea as to how this information kicks into the chemical components. How did the genetic code arise, and what is the mechanism for translating this code into protein synthesis? There are huge explanatory gaps here. Scientists have not explained how self-replicating molecules came into being. There is no certainty regarding the origin of nucleic acids, or of the process from nucleic acids to protein; no certainty regarding the origin of the cell; there are no adequate explanations of how large complex molecules, like DNA, evolved from smaller molecules. There may be 'missing link' molecules, precursors of DNA, which could have taken over the function of self-replication. Many candidates have been proposed (Shapere, 1991: 113) but none so far have achieved any form of consensus support. Concluding an inquiry into the origins of life on Earth, Jakosky says:

> It is unlikely that we will ever fully understand the origin of life on Earth. The transition from nonliving, prebiological chemistry to the existence of entities fully capable of reproducing or passing on genetic information, of changing through a 'survival-of-the-fittest' Darwinian evolution, and of metabolizing energy and using it to perform these functions, almost certainly was not a sharp transition. Rather, it must have involved a long series of steps, with only a few having been demonstrated experimentally.
>
> (1998: 109)

There are, however, three certainties: [1] life did get started; [2] it could not get started here now; and [3] we have no satisfactory models as to how it got started.

The third explanation – that life came to Earth from external sources – has several versions. One view is that diffusion of life occurred through colonization, or that living forms developed out of materials left behind by ET explorers several billions of years ago. The latter view – sometimes referred to as the 'garbage theory of life' – is unflattering, as it suggests that the evolution of complex forms of life originated in waste products similar to those left by explorers in the Antarctic or the Himalayas. There is no evidence to either support or refute the garbage theory.

Another version of the theory that life originated external to Earth is the panspermia or 'seeding' hypothesis. The term 'panspermia', which means 'seeds everywhere' was coined by the nineteenth-century Swedish physicist, Svante August Arrhenius, who held that life was seeded by micro-organisms from space. The English biologist, John Haldane, advanced a similar thesis in the early twentieth century. Philosophically the panspermia hypothesis might appear to be a weak explanation, as it only moves the question back: where did life originate from in the first place? The most obvious answer is 'somewhere where the conditions required for the emergence of life are more favourable than those on Earth'. Given the difficulty of demonstrating how life emerged on Earth this might not be such a ridiculous suggestion.

One version of the panspermia hypothesis maintains that suitable spores, chemicals, and bacteria brought life from passing comets. Fred Hoyle (Hoyle and Crick, 1981) argued that comets have been carriers of biomolecules, virus and complex cell-type bodies. The main scientific objection to Hoyle was that organic matter would not survive molecular decomposition in the radiation environment of space. However, it could be replied that organic molecules could survive this, as well as the hazards of entry into the Earth's atmosphere, if they were embedded within the ice of a large comet. Although no fossils had been found in cometary or meteoritic matter, organic molecules and amino acids had been discovered in meteorites, giving rise to speculations that life arose as a consequence of meteor bombardment. Thus Fred Hoyle and Chandra Wickramasinghe (1977: 402) argued that life on Earth originated in microbe invasion from outer space. Life on Earth, they held, began as the result of a cometary impact, and that continuous bombardment contributes to changes in living forms. In support of their hypothesis they appealed to the great epidemics throughout history. Evidence of ET life, they concluded, should be sought among the comets.

This view proved to be too outlandish for the scientific community at the time and as a result the panspermia hypothesis was discredited. A useful speculative approach could be based on an appeal to the random distribution and collision of organic molecules. Hoyle and Wickramasinghe located the origins of life in the early stages of stellar formation and argued that 'the interstellar cloud of gas and dust in which our Solar System formed continues to add biomolecules long after the early high temperature phase of the solar nebula and of the planetary material has been completed' (1978: 13). They maintained that the interstellar cloud contains the building blocks of life and consequently life on Earth was derived from these building blocks which had been prefabricated in various parts of the galaxy. Such a theory is suggestive that life could be evolving in many locations throughout the universe and Hoyle and Wickramasinghe claimed that it is virtually certain that 'similar experiments in biological assembly occurred on innumerable occasions in many other places in the Universe' (ibid.: 20).

Hoyle also argued that fundamental building blocks for life, such as the amino acids, were unlikely to form naturally in the primordial 'soup' on Earth. In 1994 astronomers found evidence of the amino acid, glycine, in an interstellar cloud called Sagittarius B2, thus supporting Hoyle and Wickramasinghe's theory that the building blocks of life occurred elsewhere. But the problem lay in finding an explanation as to how they came to Earth in large numbers. One explanation could be derived from research undertaken by Luann Becker and Jeffrey Bada of the Scripps Institute of Oceanography in San Diego. Becker and Bada have been examining the Sudbury crater in Central Ontario. This is the second largest impact crater in the world and measures 60 kilometres by 27 kilometres. Their research has uncovered a large number of buckyballs (football-shaped cages of carbon atoms) among the debris. Together with Robert Poreda of the University of Rochester, Becker and Bada have drawn the conclusion that these carbon molecules were not formed on Earth, but were formed before the solar system emerged and were swept up by a comet which eventually crashed into the Earth forming the Sudbury crater (Pool, 1996).

Critics of panspermia have drawn attention to the problem as to how microbes or elementary living particles could escape from their own planetary system and become attached to comets or other planets such as Earth. In reply it could be argued that it is theoretically possible that microbes in clouds of dust particles could be carried away by the strong repulsive force of radiation pressure from a planet's home star. This radiation is greater when a star reaches its final red giant stage and could expel dust cloud organisms into space where they would travel until trapped by the gravitational pull of a passing comet, or solar system. In this way life would be regenerated from a dying system.

Hoyle and Wickramasinghe's theory continues to gain in plausibility. In January 2000, Sandip and Sonati Chakrabarti, at the S.N. Bose National Centre for Basic Science in Calcutta, constructed a computer model to show how chemicals can evolve in a collapsing interstellar cloud, which revealed the possible large-scale production of the DNA base adenine.

Directed Panspermia

Francis Crick (1981) raised more than a few eyebrows in the scientific community when he revised the suggestion that life had been exported from extraterrestrial sources and proposed a theory of Directed Panspermia to account for the origin of life on Earth. Crick speculated that over 4 billion years ago – when the solar system was being formed – there already existed within our galaxy a very advanced civilization. It is likely, he speculated, that they discovered many potentially habitable zones which they seeded with life. The best way to do this would involve sending bacteria – as soon as the planet cooled – by means of automatic space probes. Bacteria would survive the long cold journey through space and could thrive on a young planet without oxygen.

Crick's speculation could be dismissed as science fiction, but in doing so we would lose valuable insights into the generation of new scientific ideas. It may be instructive to recapitulate the development of Crick's account of Directed Panspermia.

Crick (1981: 15) calculated that even with the help of happy accidents, such as the extinction of the dinosaurs, it would take 'roughly four billion years from soup to man'. This, of course, is based on assumptions about evolution on Earth and Crick acknowledges that there is no way of showing that this could be the case elsewhere. Crick then considered the estimated age of the universe to be about 10 billion years (opinion now appears to range between 8 and 15 billion.) This could mean that intelligent life could have evolved from scratch twice in succession. Given a billion years for the evolution of early plants, life in some parts of the galaxy could have emerged 9 billion years ago, and some 5 billion years ago it could have reached a technological level similar to our own today. This could place them 4 billion years ahead of us if they survived. In any case, it would seem that they might have had time to build a rocket containing some simple life-forms and send it to Earth. With their higher technology they could have manufactured unmanned space probes which explored the nearby stars and, within several hundred years, reported back on the suitability for life in various regions. Suppose, says Crick, they found that life is very rare. Or maybe they never found any at all, or concluded – rightly or wrongly – that they were alone, even though they found places where it could survive and adapt. Possibly, like us, they knew how difficult it is for life to spontaneously emerge. They might have known, also, that in the long run their own civilization was doomed, and very likely thought of exporting life. The distances could have been too great to set up and manage colonies. Perhaps they had tried colonizing nearby planets and, for various reasons, reached an end to colonial expansion. But this still left the possibility of sending simple forms of life to potentially hospitable planets. Men, monkeys and mice might not survive the long journey or the conditions on a host planet, which may lack an oxygenated atmosphere in its early stages. But bacteria are tremendously versatile; they can live in hot springs and barren deserts, and even survive conditions of intense radiation. In fact, Crick was only speculating about what is now considered a feasible proposal for the 'greening' or terraforming of the planet Mars.

Crick then considered rockets travelling at 1/100 speed of light, which is about ten times faster than currently known speeds. They would reach other star systems of 100 light years distance within 10,000 years. Bacteria could survive this time. Moreover, these micro-organisms could be genetically modified to assist survival and adaptation to their new environment. Directed Panspermia overcomes two objections to previous panspermia hypotheses, namely [1] the vulnerability of spores to destruction by solar radiation; and [2] the lack of a plausible natural mechanism for impelling the spore-bearing containers out of the original planet's gravitational field. An intelligently designed rocket will provide protection for the spores and guarantee their escape from the original planet.

The hypothesis of Directed Panspermia could be dismissed as a piece of speculative nonsense. It is certainly untestable. But it is important to consider some of the philosophical issues raised by Crick, especially if we note processes

84

of reasoning in the context of discovery. It is important to note that Crick was not doing quantifiable science; he was not testing or falsifying well-formed hypotheses. Directed Panspermia was not a hypothesis that could be tested as a finished research report in the context of justification; it was, nevertheless, a piece of reasoning that could be examined in the context of discovery, where the requirement for evidential support is less important than the need to indicate where new and even controversial claims might lead. The status of the actual thesis that Crick was proposing must be made clear, lest we fall into the errors made by his critics: Crick was defending the claim that Directed Panspermia is a plausible thesis, worthy of further pursuit, before it is either ruled out, adopted and possibly confirmed. To advance the case for plausibility of Directed Panspermia Crick, together with L.E. Orgel (1980: 35), appealed to the theorem of detailed cosmic reversibility: 'If we are capable of infecting an as yet lifeless extrasolar planet, then, given that the time was available, another technological society might well have infected our planet when it was still lifeless.' The authors stress that this theorem establishes a case for the possibility, not the probability, of Directed Panspermia.

Nevertheless, there are arguments in its favour. When contrasted with orthodox theories which postulated that life got started on Earth on its own – theories which have not yet been supported with satisfactory models or satisfactory experimental evidence – Directed Panspermia provides an extension of the timescale and suggests that the origin of life may require a very different environment to those where life can survive. Thus it not merely transfers the problem of the origins of life further back in time, it could resolve the mystery surrounding its origins, as the other place might be more suitable for its early development. Suppose, for example, we discover that life could never have started here, that there is some, as yet, unknown mineral or compound of crucial catalytic importance that is not found on Earth. This is not inconceivable, given the problems in finding a model for life's origin.

The claim that Directed Panspermia provides an extension of the time required for life to get started has, however, been criticized on the grounds that it violates the Copernican principle which is one of SETI's guiding beliefs. The Copernican principle states that our portion of the universe is fairly typical, and this yields the belief that if life can occur once, then in an infinite universe it can occur an infinite number of times. Paul Davies (1995) employs this argument against proponents of Directed Panspermia. There is, however, a twofold response: first, the Copernican principle has not been decisively verified, and, second, the scope of the concept 'typical' is undetermined. How typical is typical? How typical is our portion of the universe? In what respects is it typical? Mars and Earth are typical planets sharing a similar portion of the universe, but life is abundant on one and probably does not exist on the other. A consideration of Directed Panspermia may well be the way forward, thus generating alternative models to those which are restricted to analogies with conditions in the young Earth. Directed Panspermia thus offers an opportunity to generate alternative plausible hypotheses, which is an excellent example of reasoning in the context of scientific discovery.

In recent years, panspermia – if not Directed Panspermia – has gained in plausibility over the rival *de novo* theory. Perhaps the way forward for *de novo* theorists would be to abandon the assumption that complex forms of life developed as a result of a series of random chemical reactions. Paul Davies (1995) who supports explanations of life in terms of natural processes, rather than Directed Panspermia, maintains that life did not emerge as the result of accidents in the soup, but through more refined processes of self-organization which are found in chemistry and biology in ways that challenge neo-Darwinism. While Davies' theory has considerable merit, it is neither incompatible with Directed Panspermia nor strong enough to rule it out. The emergence of complex forms of life by principles of self-organization could very well be co-existent with non-natural means of disseminating life. Techniques of genetic engineering on Earth already supplement natural evolutionary processes. Crick's postulation of an advanced civilization which exports micro-organisms is by no means incompatible with a theory which appeals to more universal principles for the emergence of life.

The rebirth of the 'seeding' hypothesis

When Hoyle and Wickramasinghe first voiced their theory in the 1950s, that simple prebiotic chemicals existed in outer space and consequently 'seeded' potentially habitable planets like Earth, there were few members of the scientific community who were sympathetic. Yet during the past ten years the 'seeding hypothesis' has returned to a position of respectability. The main difference between the old panspermia hypothesis, represented by Arrhenius (and to a certain extent by Hoyle and Crick) and the new versions, is that for the latter it is not life that is being conveyed through space, but prebiotic chemicals. Nevertheless, the recognition that the essential chemistry of life exists in space and that the Earth was probably bombarded with organic chemicals in its early stages of development is a validation of Hoyle which was unthinkable several decades ago. Astronomers have discovered over 100 organic molecules in space, many of which are fairly complex. Hoyle and Wickramasinghe's thesis was given further support in 1994 when Lewis Snyder and his group of radio astronomers at the University of Illinois claimed to have found fundamental building blocks of life in a dense cloud of gas and dust near the centre of the galaxy. A report in the *New Scientist* of June 1994 indicated that Snyder *et al.* had detected 'spectral lines characteristic of many large molecules in the cloud, called Sagittarius B2, as well as glycine, the simplest of the amino acids which link up to form proteins' (Hecht, 1994: 4).

John Gribbin (1994: 66) cites estimates that soon after the Earth was formed the amount of organic matter raining down on it from space must have been as much as 10,000 tonnes each year. He concludes that the notion that life started from scratch in little warm ponds on Earth is outmoded. In fact he maintains that life on Earth began as the result of a form of 'seeding' of organic molecules as complex as amino acids. This, he claims, is a conservative interpretation of the evidence. Alfred Vidal-Madjar (1994: 413) also cites recent investigations revealing the presence of numerous amino acids that are not found in living organisms on Earth and states that the existence of complex organic compounds of extraterrestrial origins is well established. As Gribbin points out, among the advantages of the 'seeding' hypothesis is the extension of time it allows for the complex origins of life to emerge. Life can pre-date existence of the Earth, as complex chemical processes may have been underway for many million years longer than the Earth.

Molecular clouds which allegedly originate in the spiral arms of the galaxy – where stars are born – are now said to contain several complex chemicals, including biochemically important ones such as amino acids, ammonia and formaldehyde. This might indicate a level of greater chemical complexity than is believed to have existed in the early stages of planetary development. There may well be complex molecules in space ready to rain down and promulgate life in hospitable planets. One possible site is the interstellar medium where precursors such as hydrogen cyanide have been found.

Bioastronomy

The laws of physics and chemistry may apply throughout the universe but biology has been an Earthbound science. The prospect of discerning extraterrestrial life offers biology its greatest ever challenge. Life has already had an opportunity to spread throughout space: spacecraft have crashed on the Moon and have gone beyond the solar system with a probability that micro-organisms remained aboard. Who knows whether they will survive and adapt?

During the latter half of the twentieth century an entire branch of science has emerged which is based on theories concerning possible forms of ET life. Formerly called 'exobiology', but re-named 'bioastronomy', it involves the study of evolutionary lines that have not developed on Earth. It has been described as the only sub-field of biology that exists without any secure knowledge of whether it has a subject. It has been compared with theology which is concerned with a subject matter that is beyond understanding, and whose existence is disputed, yet claims, like science, to be concerned with certain truths and the methods for proving them. A statement from the Space Science Board of the USA (1962: 1079) captures the optimism of the exobiology pioneers:

> The scientific question at stake in exobiology is the most exciting, challenging and profound issue not only of the century but of the whole naturalistic movement that has characterized the history of Western thought for over three hundred years.

SETI scientists now prefer to speak of bioastronomy, emphasizing that there is a subject one can point telescopes at. Bioastronomy operates with the same Darwinian rules as terrestrial biology – mutation, natural selection and speciation – despite possible differences of form. Bioastronomers believe that the evolution of intelligence may accelerate or be restricted by various factors, and that it is selectively advantageous. Given that SETI scientists concentrate upon possible Earth-type planets, there is speculation regarding convergent paths of evolution: for example, eyes have independently evolved on at least forty Earth species, although this is not a necessary feature. However, organs of vision, if they evolve, are expected to respond to roughly similar wavelengths of light to our own.

NASA's bioastronomy programme seeks to understand the origin, evolution and distribution of life in the universe. Bioastronomers divide on the issue whether life will evolve elsewhere as it has done on Earth. Some, for example, maintain that the genetic code, responsible for the structure and make-up of all living things on Earth, evolved from a precise chemical interaction that would occur in the same conditions anywhere in the universe. According to these terms, extraterrestrial life would be broadly similar to that found on Earth. It should be noted, however, that with only slight variations in a territorial niche there could be massive differences in life-forms. Even within the same territorial niche, such as a swamp, there can be variations in living phenomena from an insect to a hippopotamus. A planet with greater gravity would favour life-forms that crawled flat, like crocodiles, or lived in water or swamps where gravity could be countered by buoyancy. The size of life-forms might not be comparable to ours, and life-spans could well be different.

Many bioastronomers are speculative, and consider possible improvements over human form that an advanced species might have. Their speculations include the prediction of larger craniums to house the larger brains usually associated with highly developed species, and possibly extra limbs to facilitate extra dexterity. Further speculations are drawn from the 'what-might-have-beens' of Earth's biological history. Suppose the dinosaurs had not become extinct some 66 million years ago, but continued to evolve while developing brain capacity and physical dexterity equivalent to *Homo sapiens* today. With their superior strength they would clearly maintain a position as the dominant species. Is it likely that contact with an ET life-form could be an encounter with the descendants of the dinosaur?

Exotic life-forms

It is frequently asked whether other life-forms on other worlds would be constructed out of the same materials as life on Earth; that is, principally of carbon, hydrogen, oxygen and nitrogen. One imaginative and very different form of life was described in Fred Hoyle's novel, *The Black Cloud*, which involved a high form of intelligent life organized in a cloud nourished by light from the stars. The most commonly discussed alternative are life-forms where silicon or boron is substituted for carbon. There is ten times more silicon than carbon in the Earth's crust, which would suggest that if it were a viable alternative it should have emerged. However, silicon is ten times less abundant than carbon throughout the solar system. This would suggest that if silicon is unsatisfactory where it is most abundant, it is less likely to be prevalent elsewhere. Moreover, there are problems in reproducing the complex storage of information necessary for the genetic code with silicon molecules. Silicon cannot form molecules complex enough to sustain life, whereas carbon can form multiple bonds with other atoms, thus allowing a vast variety of different forms of molecules. So far there has been no success in laboratory attempts to develop an 'organic chemistry' based on silicon.

A further problem with silicon is that it cannot form double bonds in the way that carbon can, which allows the latter to combine with oxygen atoms to form CO_2. Thus carbon can exist as a gas and can dissolve in water, which enables it to move through the atmosphere and the oceans. Nevertheless, in crystal form silicon can be used to make complex structures. Silicate crystals could store information and possibly reproduce their lattice structure, and even generate mutations such that new crystals could flourish. Jean Schneider, of Meudon Observatory, has proposed a crystalline form of life, but its capacity for reproduction seems limited. More complex developments seem highly improbable. If they breathed, they would breathe out silicon dioxide (sand). But could we call this breathing?

Boron, another candidate, is quite rare; it is about ten times less abundant than carbon on Earth and about one million times less than carbon in the solar system. There is also a problem with regard to the detection of either silicon- or boron-based forms of life. Thus if a radically different biochemistry emerged it would be unlikely to be detected (McDonald, 1998a: 16–17).

It would seem that the most favourable basis for life would consist of carbon compounds forming in water. Carbon is common throughout the universe, which gives it a good chance of replication. But it can only emerge in a living system on surface temperatures similar to Earth. However, it might be noted that even with this limitation there are millions of possible variations. As in the case of Earth, it would only require slightly different environmental conditions to favour very different life-forms. Does this provide credibility for the creatures of science fiction, or are there limits to biological liberalism on other worlds?

In much of science fiction there are accounts of gigantic insects and various exotic life-forms, but an appreciation of the problem of scaling reveals a limit to biological liberalism. Contrary to the beliefs of many science fiction readers, large-scale replicas cannot function if reproduced in exact proportion to small-scale originals. For example, it is often said that given an ant's ability to lift many times its own weight, were the ant scaled up to the size of an elephant its physical strength would be beyond anything imaginable. Hence a planet inhabited by giant ants could exhibit feats of engineering far beyond anything attempted on Earth. However, an understanding of scaling will soon eliminate such science fiction scenarios–along with optimistic beliefs in genetically pro-grammed chickens the size of cattle. Since Galileo, physicists, designers and architects have been aware of the problem of scaling. Galileo noted, with regard to structures having the same physical characteristics, such as shape, density or chemical composition, that weight W increases linearly with volume V, whereas strength only increases like a cross-sectional area A. Given similar structures $V \propto L^3$ and $A \propto L^2$, where L is a characteristic length or height, it can be concluded as shown in Table 3.1.

Table 3.1 Galileo's definition of weight and volume

Strength		1	1
$\dfrac{A}{V}$ $\propto$	$\dfrac{1}{L}$ $\propto$		$\dfrac{1}{W^{1/3}}$

$$\frac{\text{Strength}}{\text{Weight}} \propto \frac{A}{V} \propto \frac{1}{L} \propto \frac{1}{W^{1/3}}$$

Although smaller animals may appear stronger than larger ones, and ants appear – relatively speaking – stronger than elephants, their advantage would not survive scaling. If an ant were scaled up to the size of an elephant its weight would increase faster than its strength, ultimately causing it to collapse under its own weight. Thus the fantasies of science fiction writers of giant ants, spiders and the like, can never be realized. The giant worms of *Dune* are physically as well as biologically impossible. For similar reasons, humans cannot fly under their own muscle power, while small animals can leap proportionally greater distances than larger ones. These principles are recognized and applied by architects, designers, ship-builders and biologists and must play a profound regulatory role when considering exotic forms of extraterrestrial life.

According to a thesis developed by Roland Puccetti (1968) there are very strict limits to biological variation, and even stricter limits governing the development of intelligent extraterrestrials. As an example Puccetti cites mythical creatures such as centaurs and griffins, which could not exist anywhere because they make no evolutionary sense. The centaur has the upper half of a human body upon the quadrupedal body of a horse. With its human head we can suppose a human brain, which is hard to imagine evolving in the body of a foraging field animal. Puccetti considers the arms: early humans developed arms by crawling, climbing trees and swinging from branches, which would be impossible with a horse's body. This evolutionary trend is doomed at the outset: either the upper or lower parts would disappear or, more probably, they would never develop. The evolution of a griffin is equally implausible. The mythical griffin had the head and wings of an eagle and the body and hindquarters of a lion. Flight was clearly impossible, as it would require a massive wingspan to carry the lion's body with 'self-spined heavy hindquarters hanging down and contributing nothing to the flight' (ibid.: 89). And if it did not fly, then what use would wings be to a lion?

Despite widespread variations, evolutionary developments require that the parts contribute to the function of the whole; hence non-functional uneconomical parts are either discarded or never develop. Puccetti also appeals to a principle of convergent evolution in order to show limits on the form of possible extraterrestrials who, he argues, if they exist, would be not that dissimilar to terrestrials. For example, most predatory animals with complex nervous systems have an anterior mouth, with its sensing organs and brain relatively near, and a posterior anus. With the transfer to terrestrial surfaces there is a likelihood of greater integration of nerve centres, and legs (wheels do not evolve in living organisms) would be required for mobility, especially in a creature with a large brain which would require a heavy structure. In fact two pairs of legs are most promising for terrestrial creatures with a large brain, and this would facilitate the development of arms in the tool-manipulation phase. Convergence, argues Puccetti, is not accidental; it represents a series of common solutions 'available in the physico-chemical environment common to all planetary surfaces on which life-systems can evolve' (ibid.: 96). Thus, for example, a land predator will develop skin, not feathers or scales, and a tool-making animal will evolve fingers, rather than claws.

Although the evolutionary process is based upon random mutation and is consequently unpredictable, there are nevertheless certain convergent solutions which reveal an element of predictability. Most animals with muscles and nerves equipped for mobility are likely to develop a visual system, as the ability to see confers selective advantage. Sight has independently evolved at least three times: in vertebrates, in insects and in molluscs (squids and octopuses). Some evolutionary developments may be one-off accidents, but mechanisms such as sight are more inevitable. On these terms, extraterrestrials – if they exist – would not be greatly unlike terrestrial creatures, sharing a similar environment, similar structures and similar sense organs.

The contingency of life as we know it

SETI optimists could very well underestimate the precarious nature of continuous life. Knowledge about the Earth's history suggests that the continuance of living phenomena is precarious. The dinosaurs lasted about 150 million years, while humans have been around for less than 150,000 years. In 1980 Luis and Walter Alvarez, at the University of California at Berkeley, proposed a theory that the dinosaurs were destroyed as the result of a collision between the Earth and a large meteorite at the end of the Cretaceous period, about 66 million years ago. The original scenario implied the existence of a dust cloud, caused by the impact, which blocked photosynthesis, leaving plants and other members of the food chain to die. In a period between one and ten years most species on Earth died out. Others have contested this extinction rate, claiming that they disappeared over a period of 500,000 years, one species dying out after another. Alvarez's thesis was supported by the discovery of large amounts of a rare element, iridium, in rocks in Denmark, Italy and New Zealand. Although rare on Earth, iridium is common in meteors. However, some palaeontologists cite fossil evidence of dinosaurs in rocks below those containing iridium, which suggests that dinosaurs were not destroyed by meteor impact; others have appealed to analogies with volcanic eruptions, like Krakatoa in Indonesia, which did not wipe out all plant life in the vicinity, as evidence in favour of the tenacity of living phenomena.

In 1992 scientists at the University of Chicago added a new dimension to the catastrophe theory of extinction. Dating asteroid collisions from impact craters and checking them against known mass extinctions, they argued that during the 600-million-year history of cellular life, some 60 per cent of species extinctions may have been caused by disasters due to the impact of asteroids, comets or other extraterrestrial bodies. Nevertheless, it has been pointed out that not all extinctions have been due to impacts. It has been suggested that extinctions toward the end of the Permian period, about 250 million years ago, were the result of massive volcanic activity in Siberia.

It can also be argued that impact disasters with large comets have been minimized on Earth due to the unique positioning and masses of the planets in the solar system, thus giving life on Earth an opportunity not shared by other stellar systems. For example, George Wetherill of the Carnegie Institution of Washington, suggests that intelligent life on Earth might not have had sufficient time to evolve between impact disasters were it not for the presence of the two giant planets, Jupiter and Saturn, whose masses are respectively 318 times and 95 times that of the Earth and whose strong gravity deflects comets out of the solar system. Were it not for Jupiter's 'house-cleaning' role in the solar system, life on Earth might not have developed. Wetherill's suggestion is admittedly speculative, but is nevertheless based on a series of computer simulations of a solar system wherein these gas giants have a considerably reduced mass. Whereas catastrophic impacts take place every 100 million years on Earth, Wetherill's computer simulation which reduced Jupiter and Saturn's respective masses to a mere 15 times that of the Earth, revealed a likelihood of collision every 100,000 years, thus greatly reducing the time for intelligent life to develop (Crosswell, 1992a). If life were to be given a chance to develop in other stellar systems, the protection provided by similar gas giants to the planet Jupiter might be crucial. On the other hand, it can be argued that while gas giants such as Jupiter can deflect comets and other large masses, it can also attract them. There is growing evidence that asteroids wandering inwards from the main belt beyond Mars have been disturbed by instabilities in Jupiter.

While many recent discoveries of ex-solar planets indicate that gas giants are fairly widespread, their closeness to their parent star, which in some cases is closer than Mercury is to the Sun, calls into question the standard theory that gas giants exist in the outer regions of their stellar system, growing larger as their lengthy orbits sweep up material. If, according to some recent theories, gas giants tend to migrate inwards towards their stars, then the majority of planetary systems would not have the protection of giant sweepers, which may even become attractors, and life would be frequently extinguished before it has a chance to develop.

What can be concluded from research on catastrophes on Earth is that the development of complex life-forms has to take place within periods between life-destroying volcanic or impact events, and that for various complex forms of life to develop, a planet will have to be free of volcanic action or impacts for long periods.

It may be that the most typical form of life in the universe is some form of hyperthermophilic life, which can survive in extreme conditions, and can withstand the volcanic and impact disasters which periodically wipe out all other life forms. Jakosky (1998) maintains that hyperthermophilic life is the norm and everything else has developed from it. Given the wide range of energy sources that are known to have been utilized by early life-forms, it can be plausibly suggested that life may be widespread throughout the universe. But whether or not it has developed along the complex lines found on Earth is still a major unresolved question.

The tenacity of life

Discoveries of exotic forms of life, which survive in extremely inhospitable conditions on Earth have given rise to speculations regarding the possibility that life may be found in similar inhospitable sites throughout the solar system. These speculations not only intensify arguments in favour of more detailed searches of planets, such as Mars, but also generate hypotheses regarding the origin of life on Earth.

During the past few years many thousands of different strains of micro-organisms have been discovered surviving in intolerable conditions, as deep as 3.5 kilometres beneath the Earth's crust at temperatures of $113°C$ (Pain, 1998: 28–32). This could suggest that living organisms might also exist in other apparently inhospitable zones in the universe. Although satellite images of Antarctica have shown it to be barren like Mars, towards the end of the twentieth century it was acknowledged that Antarctica is teeming with life. Living microbial organisms have been discovered in frozen Antarctic lakes, which are probably the most inhospitable sites on Earth, thus suggesting that similar forms of life might exist on icebound planets or comets.

There is so much evidence of early life on Earth, when its surface tempera-tures were extreme, that Darwin's notion of life originating in a 'warm little pond' has to be ruled out. It is more likely that the earliest precursors of life were extremophiles, or hypothermophiles, which were unicellular bacteria that can withstand extremely high temperatures and high doses of radiation. Moreover, there are ecosystems on Earth that do not depend on photosynthesis, surviving within rocks deep in the ocean floor, relying upon combinations of oxygen and hydrogen sulphide. The implications for any search for extraterrestrial life is that the scope of the search can be extended, including deep below a planet's surface and within its rocks.

Investigations of extremophiles have generated two hypotheses concerning former life on Mars. The first view is that various forms of extremophiles survived the heavy bombardment of asteroids and comets which either eliminated or prevented the emergence of surface life on Mars. It is even possible, argues Paul Davies (1998: 24–9), that ejected rocks containing micro-organisms found their way to Earth thus enabling life to get started here. So it is possible, hints Davies, that Earth life actually started on Mars, and that we are the last Martians.

The second view, advanced by Jakosky, is more sceptical with regard to early life on Mars, as there was so little energy available from volcanism and plate tectonics – the kind of energy that might have triggered off life on Earth. Mars has a history of low volcanic activity, and has never developed plate tectonics. Many researchers believe that early cells formed within a period of 100 million years on Earth, drawing energy from chemical reactions as a result of volcanic activity, with photosynthesis occurring some 500 and 800 million years later. According to this view, deep-seated extremophiles on Mars would have less chance to surface and develop.

Intelligent life

It is a very big step from primitive life to intelligent life. It took life on Earth 3 billion years to go from a single-celled to a multiple-celled stage, and altogether it took nearly 5 billion years for intelligent life to evolve on Earth. If this is the case on other planets orbiting a stable sun in a thermally habitable zone, then we cannot expect to find intelligent ETs near stars that are less than 5 billion years old. This might, however, be contested with reference to setbacks encountered on Earth which might not have been encountered on other planets. Thus extraterrestrial life could have evolved faster on other planets if it did not have to re-evolve several times. Hence the average genesis of intelligent life throughout the universe may be less than 5 billion years.

If intelligent living systems are diffused by means of space travel, the probability of intelligent ETs would be considerably increased within the 5-billion-year period. Travelling at speeds approaching the speed of light it would take 300 million years to reach the centre of our galaxy. That, of course, would be the absolute minimum time in which life could be spread around the galaxy. But even if we multiply this figure by ten it would still be quicker to spread life by diffusion than wait for its random development. The time may be further reduced if we allow for the possibility that some advanced civilization is spreading out towards us. SETI bioastronomers maintain that once life appears, then intelligence is likely to follow and flourish as it bestows selective advantage on those endowed with it. However, this appeal to human intelligence is a very weak link in the chain of assumptions underpinning SETI, and its likelihood of being selected on other planets, it is sometimes argued, is minimal. Faculties comparable with human intelligence do not have to evolve.

George G. Simpson (1964a) has questioned the assumption that extraterrestrial life is likely to emerge along similar paths to that taken on Earth, and draws a contrast between the deterministic view of the bioastronomers and a more opportunist view. Evolution, he stresses, is not goal-directed, there has been a 'continual and extremely intricate branching' with Man as 'the end of one ultimate twig. The housefly, the dog flea, the apple tree and millions of other kinds of organisms are similarly the ends of others' (ibid.: 773). Evolution, he points out, produces numerous dead ends and extinctions without issue. If humans disappeared it is not inevitable that they would be replaced by another intelligent species. There is no reason, he argues, to assume a continuous development from primitive organisms to human intelligence. The turtle, which can be traced back some 200 million years into the Mesozoic era, reveals that life can get by without too much intelligence. In terms of Darwinian survival, the most successful species are rats and beetles. The development of our neuro-anatomy was not merely unpredictable, he says, it was ramshackle. Our brains were pieced together over millions of years. Against this it might be objected that the hardware is unimportant; that what matters is the software: our thoughts, consciousness and problem-solving abilities, which can operate with different kinds of hardware. But this argument, concludes Simpson, is unsatisfactory, as it presupposes a free-floating intelligence independent of our brains − not to mention our natural and social history. In fact, the only way that such an argument can be supported is by an appeal to some Divine programmer, a view which is not likely to find acceptance among SETI scientists, says Simpson.

When contemplating the likelihood of an extraterrestrial intelligence similar to ours the appeal to large numbers, the vastness of the universe, is of limited value. Given an almost infinite number of locations it might be suggested that somewhere an extraterrestrial intelligence has evolved with an English grammar and vocabulary. Maybe they have cockney accents as well. If they speak English, it is inevitable that they will speak it with an accent. But these possibilities are not merely rare, or of low probability; they are unlikely to the point of being downright impossible, as they would require an incredible duplication of countless events in human history, its wars and invasions, to emulate one of the languages spoken by humans. However, if they are in the habit of monitoring our radio transmissions, then they might easily acquire some of the peculiarities of a language like English.

If Simpson is correct in his assertion that extraterrestrial life is unlikely to develop along similar evolutionary lines to terrestrial life, then it would appear that if intelligence is widespread throughout the galaxy it is probably the result of diffusion. Opponents of Simpson and the neo-Darwinists might, however, respond by raising questions concerning the assumption that evolution is neither law-like, predictable nor progressive. For it could be maintained that progress is a feature of evolution, and that progress towards greater levels of complexity, which includes the development of self-consciousness and intelligence, is a law-like feature of evolutionary development. Evolutionists who shun notions like progress as unscientific may simply be confusing progress with destiny. Rigidly drawn dichotomies between 'chance' and 'necessity' continue to mislead. Evolution is neither a product of blind chance nor absolute necessity; it is a matter of probability determined by physical possibility. Moreover, SETI scientists have responded to Simpson's argument by pointing out that the emergence of a technological intelligence need not replicate all the sequences of mankind's development; in nature there are many different means of arriving at a similar goal.

Nevertheless, appeals to progress and complexity in evolution frequently rest on a highly selective range of examples. As Stephen J. Gould points out:

> When we consider that for each mode of life involving greater com-
> plexity, there probably exists an equally advantageous style based on
> greater simplicity of form (as often found in parasites, for example),
> then preferential evolution towards complexity seems unlikely *a priori*.
> Our impression that life evolves towards greater complexity is probably
> only a bias inspired by parochial focus on ourselves.
>
> (1994: 65)

Gould strongly rejects the view that human intelligence is a result of some natural evolutionary process of complexification. *Homo sapiens*, he says, could very well be 'a tiny, late-arising twig on life's enormously arborescent bush – a small bit that would almost surely not appear a second time if we could replant the bush from seed and let it grow again' (ibid.: 70).

What do we normally mean when we speak of 'intelligence'? It is difficult to define, but it should include several of the following: faculties for reasoning, creativity, inventiveness, imagination, foresight, reflection, an aptitude for learning and problem-solving and some communicative abilities. These should be employed in a manner which indicates a degree of co-ordination. In this respect it is a mistake to start from a definition of intelligence; rather, it is best to indicate what it actually does, and what it actually requires in order to function. Intelligent communication, for example, requires similar sensory equipment among members of a communicating group. This, however, is not a necessary feature among beings who share an identical environment. A desert can provide a home for humans, dogs, insects and rattlesnakes, none of which can communicate with each other. Our expectation that similar sensory receptors will be found among living beings sharing a similar terrestrial niche is dashed when we learn that the dog's sense of smell is 15 million times stronger than ours, and that, despite sharing a similar exposure to light, the rattlesnake relies on nearby wavelengths with its infrared receptors which enable it to sense its warm-blooded prey in the dark. This suggests that an environment so radically different from Earth would yield yet another range of diversity.

Although it is far from obvious what intelligence actually is, a scientific and technological intelligence of the terrestrial kind requires several essential features, including an ability to stand erect and walk upright. It also involves an ability to alter behaviour and the environment in ways that do not depend upon natural evolutionary processes. This would include the ability to use hands, stand erect, manipulate the environment – not merely adapt to it – and use fire. No non-human has ever learned how to do anything with fire other than flee from it. Human technology has developed by means of the accumulation and manipulation of information and the ability to employ a concept of 'improvement' – that is, the ability to perceive or imagine a better way of solving a problem than the existing one. Technological intelligence is clearly dependent upon certain natural facts, such as easy availability of fuel, minerals and metals. Our history is bound up with periods described as the Stone Age, Bronze Age and Iron Age. A planet where the metals are miles beneath its seas is not going to experience an Iron Age. Of course we can have intelligence without metals, such as the intelligence of dolphins, but this is not an intelligence that will invent radio telescopes. A scientific culture requires a lot of help from natural history. There is no chance of radio communication without metal, as electricity and magnetism would never be understood. That chain of research from Volta through Maxwell to Marconi would not be thinkable on a planet without easy access to metal. Life may proliferate without any possibility of achieving a technological intelligence. The sea floor temperature of 380°C and pressure of 250 atmospheres represent conditions as severe as those found on the planet Venus, yet it supports living communities (Impey, 1995). Nevertheless, these conditions would not permit the development of more advanced species and technological intelligence.

Intelligence, however, need not be linked to technology, or even linked to the development of tools and means of manipulating the environment. We can have intelligence without technology. For example, dolphins as well as many primates display abilities for abstract reasoning. We can also have technology without intelligence. Insects perform engineering feats, but are said to lack intelligence and the capacity for abstract thought. Many human societies have remained for thousands of years at the same level of technology. Many cultures in the Amazon and Africa have little history of technological development. There is something extremely parochial about the attempt to tie intelligence to the scientific and technical culture of North America and Western Europe. It may be equally parochial to link it too closely with human natural history. Our sensory apparatus – the greatest asset in a technological culture – is not merely limited to a narrow audio and visual range; it is governed by internal representations of the objects experienced, which vary in sophistication according to memories and imagination. Lacking similar internal representations, a species with similar sense organs to ours would be beyond our comprehension.

Earlier (see p. 91), Puccetti's appeal to a principle of convergence was described as a limit to biological liberalism. Puccetti also applies this principle to all forms of intelligent life, and maintains that:

intelligent extraterrestrials everywhere will resemble *Homo sapiens* to a considerable extent ... it follows inescapably from the fact that here on Earth animals and plants have independently evolved not only similar structures, but also similar biochemical systems and similar behaviour patterns as solutions to the same fundamental problems.

(Puccetti, 1968: 96)

If this is the case, then does it mean that if there are intelligent extraterrestrials they must be human? Puccetti rejects this, pointing out that they would not be members of the same species, capable of potential reproduction, with duplication of the same genetic code. The chances of that happening, he says, is as remote as the possibility that they would speak English. But they would, he insists, resemble humans in major respects.

It might be noted in support of Puccetti's contention that on Earth at least four distinct 'hominid' populations have evolved and come into contact with each other as intelligent beings (ibid.: 98). And if there are distinct human types, then there is a likelihood of some form of technology, which has flourished in isolated cultures throughout human history. Similar problems tackled with similar resources have yielded similar solutions. But what of a scientific technological intelligence? Is this subject to a principle of convergence? Or is the development of science a post-Renaissance European phenomenon? Puccetti argues that science exhibits convergent tendencies: its aim is control of the environment, and the speed at which modern science spread across the world indicates the peculiar survival power of technological intelligence, which suggests a universal value to science for any life-form that accepts it.

One of the main objections to SETI's prediction that intelligence is likely to be widespread is based on the neo-Darwinist claim that evolution is not law-like and is not predictable. Consequently, the emergence of intelligence is a matter of chance. On these terms the series of chances which led to the evolution of human-type intelligence and its survival and eventual hegemony are very unlikely to be repeated. Recently, however, this objection has been countered with an appeal to the phenomenon of self-organization which is a feature of chemistry, physics, economic systems and biology. According to Paul Davies (1995: 54), the development from simplicity to increasing complexity, from 'microbes to mind', is a law-like feature of all physical and organic systems. 'Life and consciousness are typical products of physical complexity, a product of law and not chance' (ibid.: 71). On these terms, biological self-organization, resulting in the emergence of life, and later intelligence, is likely to be widespread in similar parts of the universe. Thus given a similar pattern of law-like behaviour throughout the universe, and a tendency towards self-organized complexity, we can expect the emergence of extraterrestrial intelligence, but whether or not it is akin to ours will only be determined through contact.

It thus appears that the likelihood of widespread intelligent life throughout the universe depends on which model of evolutionary theory is adopted. The chance of widespread intelligence is remote according to Simpson and Gould's version of Darwinian theory but it is likely to be prevalent if Davies' appeal to law-like complexity is correct.

4

LIFE IN THE SOLAR SYSTEM
Terraforming and colonization

> The world is nearly all parcelled out, and what there is left of it is being divided up, conquered, and colonized. To think of these stars that you see overhead at night, these vast worlds which we can never reach. I would annex the planets if I could.
>
> (Cecil Rhodes, *Last Will and Testament*, 1902)

Introduction

This chapter evaluates the search for life in the solar system, assessing the planets and their respective satellites, the asteroids and comets, as potential sites for life. Proposals for terraforming potentially habitable sites are assessed as well as the more ambitious proposals for the eventual colonization of space.

Life in the solar system

Assumptions that extraterrestrial forms of life could exist on other sites within the solar system provided a rationale for the elaborate and extremely expensive decontamination programmes to which returning astronauts were subjected in the 1960s. It is hard to imagine a lifeless planet. On Earth life exists in almost every nook and cranny. A puddle left after rainfall will very soon be teeming with countless forms of life. In deserts, at the bottoms of oceans, deep beneath the Earth's crust, the phenomenon of life is ever present. But it is only recently that terrestrial scientists have possessed the equipment for the detection of planetary life. It is nevertheless important to maintain the distinction between conditions which are considered necessary for the emergence of life and conditions where life could survive and develop if generated elsewhere.

Broadly speaking, the solar system can be regarded as a potentially habitable zone. As yet, there are no uncontestable signs that life has been generated on other sites than Earth – and this is not absolutely certain, as life may have been generated elsewhere – but several extraterrestrial sites could plausibly be said to be capable of sustaining life. Although as a general rule planets or satellites should be in a relatively stable thermal zone, the possibility of tidal motion and the transfer of energy, due to volcanic activity from the inside to the surface, could compensate for the planet's distance from its star, and provide a suitable temperature to support life. Volcanic heat can be responsible for the formation of organic compounds, and as a source of ultra-violet radiation, as well as gases such as ammonia, hydrogen, carbon monoxide and methane. Water is made up of 90 per cent of the gases emitted by a volcano.

The Moon

The first Moon landings decisively ruled out the possibility of independent life on the Moon. After an analysis of Moon soil revealed no signs of life, the quarantine regulations for returning astronauts were lifted. However, strepto-coccus bacteria (found in a camera that had been left by the Surveyor 3 probe in 1966) had survived, and were brought back by the Apollo 12 mission in 1969 (Breuer, 1982: 65).

Long after the first lunar landings revealed the Moon to be a cold, barren, inhospitable place, there were some who still claimed that it was inhabited. Don Wilson (1975, 1979) argued that it was hollow, having been converted into a satellite by extraterrestrials whose own ship was damaged. This theory might survive a brief Moon landing but not the data from moonquake recorders left there by Apollo astronauts which are incompatible with the notion of a hollow Moon. The Moon's mass is too small to retain an atmosphere, but under suitable protective conditions life could survive.

One school of thought (Zey, 1994) sees settlements on the Moon as early stages in human colonization of the solar system and much of interstellar space. Colonizing the Moon has long featured in science fiction but it is now seriously considered by the USA, Japan and Russia, who envisage a permanent presence of people there early in the twenty-first century. The Moon would provide a base for future expeditions, a nearby testing ground for space habitats, and a possible fusion energy source; for the Moon has a large amount of helium 3, the proposed basic source for fusion power, in its soil. This could be sent back to Earth at low cost. According to Professor Hiromu Momote, of the National Institute of Fusion Science in Nagoya, Japan, the Moon's abundance of helium 3 – which is not found on Earth – will make it possible to build cheap, simple and clean fusion machines, which may be able to power spacecraft. A Moon colony would provide a superior location for astronomy, low gravity hospital care, and an attractive venue for commerce and tourism.

Further support for proposals for a Moon colony followed the interpretation of data from the Clementine probe of 1994 which recorded an unusual radar reflection near the lunar South Pole. Scientists interpreted this as evidence of ice water. Normally ice arrives on the Moon as a result of cometary impact and it is usually dispersed after exposure to UV radiation which converts it into atoms of oxygen and hydrogen. Sceptics have rejected the suggestion that there may be ice on the Moon on the grounds that similar reflections are found in craters that are not in permanent shadow and could not harbour ice. However, if the interpretation of the Clementine data is correct, then ice water may persist in deep craters shielded from solar UV light. Perhaps there is enough to supply a human colony, which would suggest that the Moon is a potentially habitable place. In January 1998, NASA's Lunar Prospector mission was launched, carrying equipment to measure the amount of hydrogen on the Moon's surface. An excess of hydrogen in the Moon's polar regions would be very suggestive of the presence of water-ice.

It was announced on 5 March 1998 that the Lunar Prospector had actually detected water-ice at the North and South Poles of the Moon. Estimates of the amount of water ranged from 11 million to 330 million tonnes (*The Times*, 6 March 1998: 1). If the more generous estimates are correct, then a human settlement might well be supported, and the possibility of life on the Moon could yet again be raised. However, other reports derived from the Lunar Prospector suggest that ice is in very low concentrations spread widely over the polar regions, amounting to no more than a few crystals in any given location.

Mars

The planet Mars has been the most widely discussed possible site for extra-terrestrial life, as it more closely resembles the Earth than any other planet. As late as the 1960s it was still possible to believe that Mars was a bearer of life, as its variations in light and dark markings were believed to be the response of plants to seasonal change. These light and dark markings are now held to be caused by wind-blown dust on a dry and frozen planet. The Martian temperature is -60°C. The Martian North Pole consists of water-ice and its South Pole consists of mainly carbon dioxide ice. While its South Pole temperature could drop to -130°C, temperatures at -15°C at the equator are possible. The Martian atmosphere is composed of carbon dioxide and a little nitrogen. It is widely believed that Mars was once a site of great physical activity, with large volcanoes and vast rivers, although it might be argued that Mars was never wet, and its valleys were cut out by flowing glacial ice rather than river networks. Mariner 9 surveyed the whole surface with a resolution of one kilometre, but no artefact or large plant was identified on Mars' barren crater-covered surface. Mars was revealed to be a relatively inhospitable place, an observation which reduced expectations of life being found elsewhere. If there is no life on Mars, where conditions are more similar to Earth than other planets in the solar system, then expectations of the discovery of life elsewhere might be reduced.

No signs of life were conclusively reported after the two Viking landings in 1976, although evidence from them has not been strong enough to completely eliminate the possibility that life, of some kind, persists there. Viking I probe landed on 20 July 1976 at Chryse Planitia, and Viking II at Utopia Planitia on 3 September 1976. The Viking Mars experiments initiated very sophisticated experiments in the search for extraterrestrial life. They revealed that the atmosphere contains 2.7 per cent nitrogen, an essential requirement for life, although when surface samples were analysed in a search for organic compounds, at sites several thousand kilometres apart, they revealed no signs of life. This may suggest that the chemistry of the atmosphere rules out organic development.

There were three Viking experiments (Smith, 1989). The first involved two cameras mounted on the lander to take photographs of any large plant or life-form in the vicinity. None was recorded. The second experiment involved a gas chromatograph/mass spectrometer which searched for organic molecules in the soil. No signs of life or its precursor was found. There was no evidence of existing or fossil life. However, the dried-out Martian lake beds might be a good place to look for fossils, and plans are underway for future landings in the vicinity of Martian lakes. There may be fossils in the sedimentary deposits of these ancient lakes that existed over 3.8 billion years ago. Organic material has been found in meteorites and Mars has been bombarded by them, so it might be concluded that something on Mars destroys organic molecules – the most widely accepted cause is held to be the ultraviolet flux from the Sun.

The third Viking experiment consisted of tests on samples of Martian soil to look for metabolic processes like those used by bacteria, green plants and animals. This search also proved inconclusive. The main reason given for the absence of life is that Mars lacks oceans of water in full view of the Sun.

Despite the inconclusive Viking experiments the case against life on Mars has not been proved beyond doubt. Biologists are familiar with exotic examples where life has been developed and sustained in extremely hostile environments. Life may exist in isolated 'oases', near the equator or in deep depressions. It may be that deeper soil samples, or samples taken from nearer the Martian ice-caps, may reveal organic compounds. If there is life on Mars, it is most likely to be found in the form of bacteria buried deep in the planet's permafrost, or lichens growing within rocks (Kiernan, 1994). It has also been alleged that there are water springs which may harbour life in some remote parts of the planet, and observations from the Mars Global Surveyor lent support to the belief that the Martian polar caps may be covering water lakes. The Hubble space telescope has shown that there is water-ice on Mars, not only in the polar regions but on high altitudes and in the atmosphere. But conditions are not believed to be stable enough for the survival of liquid water on the Martian surface.

One current view is that the surface itself could be covered with residual chemical signs of life which previous missions, such as the Viking Landers, were not equipped to detect. Since Viking it has been assumed that the planet's exposure to the Sun's ultraviolet rays has oxidized organic molecules on its surface. But several scientists have suggested that ultraviolet radiation need not destroy all organic molecules as the hydroxyl radicals may react with some organic molecules to form stable compounds and carbolic acids (Knight, 2000: 11). The Viking Landers were not equipped to detect carbolic acids, but future missions could be designed to bring the untreated sample of soil back to Earth where more sophisticated experiments could determine the presence or absence of organic compounds.

Clues regarding the kind of life-forms to seek could be found by searching for life-forms in the Earth's colder regions where conditions are more similar to Mars, although these clues would not count as evidence. Lichens have been found within Antarctic rocks, which protect them from the cold and absorb water. Bacteria have also been found in the Siberian permafrost, proving that life can exist in extreme circumstances (Kiernan, 1994). One of the most extreme examples of life is the discovery of terrestrial bacteria which live hundreds of metres below the ground, existing on a diet of rock and water and without any energy input from the Sun. Instead they derive their energy from chemical reactions between rock and water. These bacteria were found in deep aquifers near the Columbia River by Todd Stevens and Jim McKinley of the US Government's Pacific Northwest Laboratory in Richmond, Washington. They called them a 'subsurface lithoautotrophic microbial ecosystem', or SLIME. This form of life could be present on Mars, where SLIME could survive if they are there. The problem is that we have no idea how they originated on Earth. But if they are precursors of more complex forms of life there would be a strong case for a search for Martian SLIME (O'Hanlon, 1995: 19).

In August 1996, the belief that Mars was once inhabited by primitive life-forms was revived when NASA scientists announced in a blaze of fanfare that the meteorite ALH84001 contained microscopic fossils. NASA researchers had examined this meteorite, which weighed 1.9 kilograms and was collected in Antarctica. It is said to have been chipped off Mars following a collision with an asteroid or comet. About 13,000 years ago this rock came close enough to be drawn into the Earth's gravitational field. After detailed studies with high-powered microscopes and laser-based techniques of chemical analysis, it was claimed that certain markings resembled the outlines of tiny cells, and there were tiny beads of magnetite similar to those excreted by terrestrial bacteria.

The announcement that primitive life once existed on Mars, together with speculation that some forms of life may still survive there, stimulated a flurry of interest and support for Martian voyages. US President Clinton said: 'I am determined that the American space programme will put its full intellectual power and technological prowess behind the search for further evidence of life on Mars.' Announcements of planned missions were also made by Japan and Russia.

Sceptics were quick in their denunciations, suggesting terrestrial contamination or inorganic processes that mimicked signs of life. In reply it was argued that after crashing in the Antarctic the rock was swept along ice-floes, perfectly refrigerated and uncontaminated, until it was collected by scientists in 1984. However, while early analysis seemed to suggest that the fossil findings originated as early Martian life-forms, the issue was far from decisive and a very confusing picture emerged. Some opponents argued that the meteorite had been contaminated during the examination process or by components from the

Antarctic ice (Holmes, 1996: 4). In reply it was argued that this was unlikely as the meteorite handling facility at the Johnson Space Center has been dealing with meteorite samples for thirty years and has developed very tight procedures to minimize possible contamination (Jakosky, 1998: 146). It was also unlikely to have been contaminated during the 13,000 years it rested in the Antarctic ice. Other meteors have been collected from the Antarctic and tested in the same way as ALH84001 with no evidence of contamination.

There were disputes over the age of the fossils and doubts were expressed whether bacteria of the alleged size could actually contain genetic material. Moreover, no evidence of cellular structures had been detected in the fossils prior to the announcement (see Kiernan *et al.*, 1996: 4–5). One sceptic, Dr Monica Grady, Curator of Meteorites at the Natural History Museum, London, who has examined chunks of ALH84001, told the *Observer* (11 August 1996: 20): 'I am completely unconvinced there is any evidence on this meteorite to support the idea that life once existed on Mars.' Sceptics argued that the evidence presented did not conclusively rule out the thesis that what had been observed in the meteorite could be explained with reference to inorganic processes. Other sceptics have argued that the temperature in which the alleged microfossils were formed was probably too hot for any micro-organism to survive. The sceptics maintained that the tiny egg-shaped structures were too small to represent living forms, although claims have been made that some forms of terrestrial bacteria are almost as small.

One intriguing feature is that while the Martian samples are smaller than any known bacteria, their overall shape is similar and very unlike non-biological structures. A further hint that living forms the size of the Martian fossils exist comes from Western Australia, where the geologist, Philippa Unwin, and her colleagues claim to have detected living forms in sandstone deep beneath the seabed in Western Australia. These candidates are called 'nanobes' and measure between 20 to 150 nanometres in diameter. Although their credentials as living organs have not been determined, if nanobes are living structures they would add support to the claim that the Martian meteorite actually contains fossilized bacteria, as some of those fossils were merely 20 nanometres across (Dayton, 1999: 13).

So far the evidence is not strong enough to rule out either the claims of biological structures or non-biological structures. However, claims in favour of the existence of primitive life continued throughout 1996. In November British scientists from the Open University and the Natural History Museum announced that they had discovered further signs of microbal life in the Martian meteorite, 79001, that was formed 180 million years ago.

An outright dismissal of the hypothesis that there is or was life on Mars may be premature. Questions still remain open about life in the distant past, which could be resolved affirmatively by the discovery of fossils in the great ice cliffs. The debate will not be conclusively settled until a widespread search has been conducted, which will include samples taken from beneath the Martian surface.

Now that the search for life on Mars has been regenerated, a series of NASA spaceflights have been proposed. In November 1996 the Mars Global Surveyor set out to bring back samples of the Martian surface in the year 2005. In September 1997, Global Surveyor began its orbit of Mars. This project reveals assumptions concerning the probability of Martian life, as samples of the rock will be subject to quarantine, according to NASA regulations, and will continue to be so when the space probe returns in the year 2005. Quarantine will, however, add many millions of dollars to the cost of missions (Kiernan, 1997: 6). Nevertheless, Global Surveyor is expected to generate as much data as all previous Mars missions combined. Since March 1998, it has been obtaining photographs with a resolution of a few metres and searching for potential future landing sites, such as ancient lake beds which may also be useful places to search for signs of former life. Despite the ill-fated Mars Polar Lander in December 1999, further missions are planned. The European Space Agency intends to launch the Mars Express Orbiter mission in 2003, with a British lander, Beagle 2, which will obtain soil samples, and further missions are planned for 2005 and 2007, which will obtain soil samples and deploy robots.

These surveys will look for evidence of life, and evidence of conditions which could have supported life. Geological evidence suggests that ancient Mars – about three and a half billion years ago – once enjoyed a warmer, wetter climate, with liquid water at or near the surface. There is evidence of water erosion in networks of valleys and flood channels, hot springs associated with volcanic activity, and erosion of the sides of impact craters. However, high resolution images from the Mars Global Surveyor in 1999 have suggested that what have appeared to be dried-up river beds are more likely to be channels caused by an upsurge of water from within the planet in a one-off flood. These data are incompatible with theories which are suggestive of rivers fed by rainfall and a climate roughly similar to Earth. Rivers fed by rainfall have a distinctive pattern of tributaries, and over long periods of time the course of a river may change. But this does not seem to be observable on Mars, which suggests that apart from one massive flooding, Mars has been dry throughout its history. If this is the case then theories regarding the former existence of complex living systems will be weakened. But evidence could be sought of former hot volcanic springs which could have supported primitive forms of bacteria. If there were hot springs, then a search in their vicinity would be an ideal place. Images from the Mars Pathfinder project in 1997 are indicative of a former water-rich environment in the Aires Vallis region. Sand dunes and possible water-worn rock conglomerates suggest that water once flowed through this valley. In the next few years the planned missions may well incorporate a search for signs of life that the Viking missions may have missed (Jakosky, 1996: 39–42).

The more exotic contemporary claims regarding life on Mars centre on interpretations of a number of anomalies in a region of Mars known as Cydonia. One of the Viking photographs (frame 35A72) taken from an altitude of 1,000 miles with relatively poor resolution, has been interpreted as being representative of a massive Sphinx-like face, about 1.6 miles long, 1.2 miles wide and just under 2,600 feet high. Other anomalies which have been cited in this region include pyramids, an ancient city and a fort. One alleged five-sided pyramid has naturally invited comparisons with Egypt's Great Pyramid (see Bauval and Hancock, 1996: 40–2). The argument turns on whether these photographs represent artificial or natural phenomena. In a paper presented to the Department of Physics and Astronomy at the University of Leicester, Dr Tony Cook dismissed claims that there is evidence of an ancient civilization on Mars which is often claimed with reference to the famous 'face' formation. Cook suggested that this formation is similar to the mesas in the Arizona desert (Mobberley, 1994). This is the view echoed by NASA, who have been accused of concealing the truth about a former Martian civilization. The matter appears to have been resolved by the Mars Global Surveyor which went into orbit around Mars in September 1997, and took photographs of the surface and the surrounding terrain including the Cydonia region. The Surveyor employed a high precision camera which was more than ten times as sharp as the one which took the original Viking pictures. In April 1998, Surveyor released photographs which indicated that the 'face' was merely a hummocky hill with no facial features or signs of engineering at all. The Martian 'face' episode is a good example of the avoidance of Occam's razor, which requires that plausible scientific explanations are examined before more exotic claims are made.

The two small Martian moons, Deimos and Phobos, were once speculated to have been spacecraft abandoned by the Martians, but early Mariner 9 photographs, which facilitated their mapping, dispelled this idea.

Venus

The planet Venus is close to the size of the Earth; its mass is about 82 per cent of the Earth's. It has turbulent atmospheric conditions, rains of sulphuric acid and suffers from a runaway greenhouse effect caused by its dense carbon dioxide atmosphere. Its surface temperature is $475°C$, although there are protected regions with lower temperatures. There is no water there, and if there ever was any, it has long since evaporated. The atmosphere consists largely of carbon dioxide (about 97 per cent carbon dioxide with 1–2 per cent of carbon monoxide), with only traces of oxygen. Its surface pressure is one hundred times that of the Earth's at sea level and the planet is covered by a sulphuric acid cloud. Infrared radiation is trapped by the atmosphere. It is an inhospitable place for life to emerge and offers no opportunity for the survival of life-forms generated elsewhere.

Has life ever existed on Venus? There is no evidence, but if there was a period with less CO_2 and lower temperatures, possibly about 4 billion years ago, life might have developed. In fact, life might have developed on either Earth, Mars or Venus, and might even have been transferred by meteors from one planet to the other (Jakosky, 1998: 190).

Mercury

Mercury is too small to retain an atmosphere and its temperatures are too extreme to sustain life. A Mercury 'day' is 176 terrestrial days. Its hottest temperature, when close to perihelion, could be $430^{\circ}C$, falling to $-170^{\circ}C$ in its 'night' side. Its surface is dominated by impact craters, it has low surface gravity and an absence of water. The surface of Mercury has not shown any geological evolution for 3.5 billion years. It is the least hospitable of the inner planets.

Jupiter

The main problem with Jupiter is its size. Its surface gravity is 2.6 times as strong as Earth's, which completely rules out the development of complex forms of life. Its atmosphere is composed mainly of hydrogen and helium with significant amounts of methane and ammonia. Organic compounds can be formed in such a mixture, a claim which has led to speculation of life-forms akin to jellyfish floating around in the more favourable temperature zones. The Galileo mission, which arrived at Jupiter on 7 December 1995, revealed details of Jupiter and its moons. It revealed the presence of water in the top layers of Jupiter's clouds, which causes thunderstorms. Carl Sagan, in his book *Cosmos*, speculates on the form of Jovian life and imagines living balloons which can float at an appropriate level at which it is possible to survive. But this is unlikely. Microbes which can survive under severe pressure could exist there, were it not for other inhospitable features of the planet.

Jupiter has organic molecules within its atmosphere and there may be liquid water within its clouds, causing expectations of life within its atmosphere. Several years ago this speculation would have been ruled out as absurd. But towards the end of the twentieth century many species on Earth were discovered to be surviving in conditions which would formerly have been regarded as inimical to life. So, perhaps, organic development is possible in the Jovian atmosphere. However, it is very unlikely that Jovian life would emerge out of the prebiotic soup of warm liquid water, as species large enough for reproduction would be drawn by gravitation into the deeper, hotter regions of the Jovian atmosphere, where organic molecules would decompose.

Beneath the Jovian atmosphere there is no surface, either solid or liquid. Jupiter is also extremely hot; it radiates twice as much heat as it receives from the Sun. This is because heat energy is produced by the effect of gravitational contraction. There are, however, regions where temperatures are around $27^{\circ}C$ but any life-precursing molecules are unlikely to have developed as the strong winds would carry them into regions where they would be destroyed by intense heat as high as $1100^{\circ}C$.

The outer planets have more satellites than the terrestrial planets. There are 16 Jovian satellites, but many are small. These satellites, like planets around the Sun, receive a considerable amount of energy from Jupiter, which might be regarded as a mini-stellar system. The space shuttle Atlantis has undertaken observations of the Jovian satellites since July 1995. The four largest are the size of terrestrial planets and are now regarded as planets in their own right, but none have appreciable atmospheres, and they appear to be inhospitable. Nevertheless they reveal astonishing diversity in their surfaces which suggests different evolutionary development. The outermost three, Europa, Ganymede and Callisto, are icebound and airless. In 1997 the Galileo spacecraft identified molecules containing carbon and nitrogen on the surfaces of Ganymede and Callisto. Data from the Galileo spacecraft have been interpreted as radio waves emanating from Ganymede. Io has active volcanoes, is hot and without water, but it is the only one with an atmosphere, although it is almost exclusively composed of sulphur dioxide ejected from the volcanoes. This volcanic heat source could maintain life, although it has eliminated water, which means that it is unlikely to produce hydrothermal conditions for life.

The Voyager probes in 1979 revealed evidence of massive meteoric impacts on Callisto, which is pockmarked with ice craters and is probably the most cratered object in the solar system. This satellite is marginally smaller than the planet Mercury and recent data from the Galileo probe suggests that it has an internal liquid water ocean. Europa and Ganymede have distorted orbits with resulting tidal patterns which could generate heat. There is a possibility that a liquid water ocean exists under the icy surface of Europa. Although its surface temperature is -145°C, and there is no active volcanism, released images from the space probe Galileo, in August 1996, suggested that the ice crust might not be very thick, and underneath may be oceans of liquid water, heated by Europa's core or the tidal forces caused by Jupiter's gravitational pull, which flood the outer surface from time to time. Jupiter's mass is capable of raising tides in excess of 30 metres each day of 85 hours. So the main question to be resolved is: how thick is Europa's icy surface? According to one school of thought, the ice is extremely thick, between 10 and 15 kilometres deep. Another standpoint appeals to close-up photographs of Europa from the Galileo probe which indicate regions where the water has actually risen through cracks in the ice, thus supporting the theory that beneath the ice there is warm liquid water. If the surface of the ice is thin with several cracks in it, then chemicals could find their way into the ocean and signs of life might be observed near the surface. It is believed that this liquid ocean could support microbial life and even larger life-forms. Limited analogies may be drawn with life in icy conditions on Earth. But if Europa is deeply sealed with ice, such that oxygenated chemicals of the sort that maintain sea creatures in deep sea vents on Earth would not penetrate down from the surface, then Europa would not be a very likely candidate for evidence of life. Predictions that life could develop on Europa suffered a setback in 1999 when sulphuric acid was detected on the planet's surface. However, some scientists saw this in a favourable light, as sulphuric oxidants are energy sources for life on Earth.

Europa is roughly the same size as Earth's Moon and is a good candidate for exploration. The Europa Observer, scheduled for launch in 2003, will use long-range equipment similar to that used to observe hidden lakes beneath the Antarctic ice sheet in order to penetrate the layer of ice which covers Europa. This radar survey will be followed by a robotic lander which can cut through the ice to enable a more detailed survey.

Saturn

Saturn is smaller, but broadly speaking similar to Jupiter, consisting of about 80 per cent of Jupiter's diameter. Like Jupiter, Saturn is a large rotating gas planet consisting mainly of hydrogen with no surface but with a probability of a hot interior. Saturn has stronger winds than Jupiter, but lower temperatures because of its distance from the Sun. The outer planets appear to be too far away from the Sun to provide the environmental conditions that would support life. However, several of the planetary satellites appear to have some of the environmental conditions suitable for life, such as the presence of liquid water and access to relevant biogenic elements. Saturn has at least seven satellites with diameters between 400 and 1,500 kilometres, the largest of which is Titan, which is about half the size of the Earth and often regarded as a planet in its own right. There are fourteen other satellites which are very small, with diameters less than 300 kilometres, and little is known of them.

In size, Titan is between Mercury and Mars. It possesses an atmosphere consisting primarily of nitrogen, which is thick and suggestive of greenhouse warming. It has an icy surface where water-ice is most abundant and there are oceans of liquid methane, liquid hydrocarbons and a single giant continent the size of Australia. Its temperature is -170°C. It may be a site for prebiotic chemistry. Substantial organic chemistry has already occurred on Titan, of the sort that has been associated with models of prebiotic Earth by Miller and Urey. Suggestions have been made that Titan could therefore be a model for the origins of life on Earth. However, if life on Earth did not arise this way, and was brought about from migration, meteors, various exotic modes of dissemination, or in fact developed out of hypothermal systems wholly dissimilar to that depicted by Miller and Urey, then Titan might not be such a useful laboratory model for the origins of life on Earth. A model, after all, is only meaningful insofar as it is abundantly clear what it is a model of.

At present Titan does not have a biosphere, but with warming it could be a potential site for life. NASA and the European Space Agency launched a spacecraft, Cassini, in October 1997, which will begin its four-year exploration of Saturn's moon system in June 2004, but will concentrate primarily on Titan.

It is the destiny of the Sun to become a red giant star in about 6 billion years time. This process will engulf the inner planets and will incinerate the Earth, destroying all life there. But frozen Titan will be warmed to a surface temperature of -70°C, which will create oceans of water mixed with ammonia. Ralph Lorenz and his colleagues at the University of Arizona's Lunar and Planetary Laboratory, Tuscan, predict that this process could give rise to life (Hecht, 1997: 28). The Sun's red giant phase will last for several hundred million years, which is longer than it took for life to evolve on Earth. But Titan might not only be a cradle for life; it could well become a refuge for those who are fortunate enough to escape from Earth. If Lorenz's hypothesis is correct, then other red giant suns could support life on former frozen planets.

Uranus

Uranus is twice as far as Saturn from the Sun, and has an 84-year orbit. It has a substantial atmosphere, consisting mainly of molecular hydrogen, and possibly ammonia or methane, but its surface is hidden by clouds and the density of its atmosphere. It is a large gaseous planet which is an extremely unlikely site for living phenomena, although it is always possible that the gas giants, Uranus and Neptune, could support micro-organisms, but such thoughts are highly speculative. Uranus has seventeen known satellites, the last two discovered in October 1997. They range in size from diameters of 50 to 100 metres, and probably consist of icy and rocky materials, but are too small to retain an atmosphere.

Neptune

The planet Neptune is a gas giant similar to Uranus. Like the other gas planets, Jupiter, Uranus and Saturn, its atmosphere consists largely of hydrogen with traces of helium and other chemicals and the planet produces considerable internal heat. It has no surface beneath its atmosphere. Neptune has two satellites and possibly a third. The two known satellites are Triton and Nereid, so named after the servants of Neptune. Triton is volcanic and may possess an atmosphere of nitrogen and methane. It is extremely cold with a temperature of -223°C, and is a very inhospitable place.

Pluto

The planetary status of Pluto has been questioned: it is very small and more typical of the medium-sized satellites of the outer planets. One theory is that it is a planetesimal, left over from the creation of the solar system. Another theory suggests that it is an escaped satellite of Neptune. Spectroscopic observations revealed in 1976 that it has an icy composition, with frozen methane, and its temperature is about -223°C. Its desolate frozen landscape is pock-marked with craters.

The asteroids

There has been speculation that asteroids could sustain colonies from an advanced technological culture (Papagiannis, 1978) but only exploration would settle this matter decisively. It has been argued that some of life's precursors are to be found in the asteroids as several meteorites bearing organic compounds may come from the asteroids.

Comets

There are about 600 comets in the solar system; their total mass is only about 0.1 the mass of the Earth so they are not likely to retain an atmosphere. However, their total surface area is 1,000 times greater than the size of the Earth. As they contain carbon, nitrogen, oxygen and hydrogen, they could provide habitable sites for some kinds of life. Freeman J. Dyson (1973) argued that too much attention has been given to searches for planetary life and not enough to comets. The home for really big technology, associated with supercivilizations (or 'Dyson civilizations'), is more likely to be on comets, argues Dyson. Whereas planets may be important for the generation of life, supercivilizations may have outgrown them. Now if comets are distributed throughout the galaxy, some of them could be habitable oases, which could shorten the distance between contact points. If such an idea is plausible it would certainly encourage a re-assessment of ideas about ET interstellar travel. Dyson also suggests that his hypothesis can be tested by searching for high levels of infrared radiation in the vicinity of comets. However, it is always possible that very advanced civilizations might not use up so much energy and their presence on comets would not be detectable.

Colonizing the solar system

Space travel, it has been argued, is a now or never affair (Breuer, 1982), as we are using up the necessary material resources. There are proposals to mine asteroids. Space Dev is an American exploration company that wants to stake its claim to an asteroid, which is intended to be mined for its minerals when the technology becomes available (Kleiner, 1997: 18). This raises legal problems, as international treaties prevent nations from staking a claim to heavenly bodies. But the status of private companies is not that clear in this field. It has been pointed out that in about 200 years there will not be the fossil and metallic raw material for spaceships and space stations (Breuer, 1982: 256). If work is not undertaken soon to extract material from the Moon, it will be too late to exploit the Moon or the asteroids, as there will not be the material left on Earth to create the rockets capable of lifting a minimum of 2,000 tonnes of implements and a nuclear reactor to the Moon. The distances within the solar system are daunting, not only in terms of material resources, but in terms of the psychological problems encountered in long periods in space.

Mars

Mars is 124 million miles away; it took the Viking spacecraft just over a year to reach it in 1976. On the best estimates, relying upon major improvements in technology, a return journey to Mars would take between one and three years. Nevertheless, the technology now exists for a manned expedition to Mars; what is required is the political motivation. Attempts to generate and maintain interest in the exploration of Mars have been conducted by the Mars Society, who have designed the Martian flag, which is a tricolour, of red, green and blue vertical stripes. Critics have accused the flag design of anthropocentrism, as the colours represent those sensed by human eyes, whereas non-human life-forms are not likely to have the same sensors, which suggests that the same anthropocentrism that dominated nineteenth-century European colonialism is prevalent in the twenty-first century. In reply, it is argued that the three colours, red, green and blue, have a Martian basis, as they represent the three stages of terraforming the planet, where the red signifies its present state, the Martian rock; green represents plant life; and blue represents liquid water. The current scientific goal, however, is to obtain a detailed map of Mars; search for traces of water and signs of former life; and prepare for eventual human exploration.

Proposals for colonizing planets fall under two categories: the first involves the construction of biospheres within which an artificial Earth environment is maintained inside a protective shield to protect its inhabitants from hostile contingencies; the second proposal is more dramatic, involving terraforming – transforming into an Earth-like environment – the planet itself.

Between September 1991 and September 1993 an eight-strong team of men and women survived in Biosphere 2 – a sealed glass and steel structure set on 1.2 hectares of desert in Arizona. Biosphere 2 (Biosphere 1 is the Earth itself) was a closed ecological system, a prototype for eventual life on other planets, particularly Mars. The team had to overcome psychological problems as well as the physical problems of survival, and actually emerged still on speaking terms with each other as well as being physically fit, despite loss of weight due to problems caused by the project's periodic food crises. Critics pointed out that the hype took over from the science. The project was not strictly self-sufficient; it required supplementary oxygen on at least two occasions. When levels of carbon dioxide became too high, crop harvests were so poor that the inhabitants were short of energy. Moreover, the isolation, which they would have experienced on Mars, was minimized by visitors who were able to communicate from special rooms where they could see and hear each other by means of video cameras. This, of course, could be arranged on Mars. What could not be arranged is the 200,000 tourists who visited the structure each year (Veggeberg, 1993).

Nevertheless, plans for Biosphere 3 are underway. The Mars Society envisages a simulated Mars station on Devon Island in the Canadian Arctic. This island is cold and dry, covered with rocky ridges and meteor impact craters, and generally resembles the Martian environment with one exception: the Martian atmosphere is one hundred times thinner than Earth's. This research station, at an estimated cost of US$1 million, will simulate the kind of self-contained and isolated environment that will be encountered by the first human explorers on Mars.

Plans for the actual transformation of planets like Mars are extremely ambitious. The first phase in plans to make Mars fit for humans is to raise the Martian temperature, which is more ambitious than melting Antarctica. This strategy is geared to the creation of a greenhouse effect. The temperature would have to be raised from -60°C to 0°C, which would allow liquid water to settle on the surface. According to one scheme, human settlements or biospheres would be installed on Mars which would house engineers and workers who would build factories on Mars to churn out tonnes of greenhouse gases each year, thus trapping heat in the atmosphere. These would be the very same gases which are destroying the ozone layer on Earth. This would be a big project, employing tens of thousands of workers, but chlorofluorocarbons (CFCs) are composed of carbon, fluorine and chlorine, which are known to exist on the Martian surface. It may only require a little initial effort from humans; once a greenhouse effect is underway other greenhouse gases, such as CO_2, would be released, thus contributing to the runaway effect that has been predicted from studies of models of the Earth's climate.

Among the schemes for the greening of Mars are proposals to melt the northern polar ice-cap by scattering carbon black pigment over its surface to enable the solar rays to be absorbed, rather than reflected. Dark soot could be mined from two small moons orbiting Mars. Other proposals include the erection of giant mirrors in space to re-direct the heat of the Sun. This melting would then create a greenhouse effect in which carbon dioxide and water vapour would be released into the atmosphere to raise the surface temperature. Water could be imported from the asteroids if there is not enough native water. Moreover, there are many icy objects of considerable size beyond the planet Neptune. If their orbits were tilted towards Neptune for gravity assistance and then aimed at Mars, they might provide an adequate supply of water for human settlements (Johansson, 1998: 53). According to estimates (Nadis, 1994; Zey, 1994), after about 200 years a wetter, warmer Mars would possess a thicker atmosphere, more tolerable for humans.

The second phase is estimated to take a great deal longer, about 100,000 years (Nadis, 1994), and would involve the production of oxygen of sufficient quantity for humans to breathe on the planet. Thus, when the planet has warmed, certain organisms – for example, the blue-green algae found on Earth – could be planted. At present Martian soil lacks oxygen, nitrogen and phosphorous. But the planting of lichens and blue-green algae, if successful, would help to produce oxygen. Once underway, genetically altered plants and self-replicating micro-organisms could be added in order to speed up the process. As the atmosphere becomes thicker the planet would become warmed. Carbon dioxide, nitrogen and water would seep from the crust. Lakes and small oceans would appear. Rocks could be mined for iron oxide which could be heated to give off oxygen. Within several decades there might be soil. But the whole operation could take many centuries. For example, lichens have a very slow growth rate and are not suitable for fast oxygen production. If blue-green algae covered one quarter of the Martian surface it would take about 7,000 years to produce 5mb of oxygen; the minimum for human breathing at this rate would take 140,000 years (Smith, 1989: 124). The best solution to this problem lies in further developments in genetically altered plants which could speed up oxygen production.

There is a possibility that bacteriological research could contribute to the terraforming of Mars. Following the claim that bacteria may have once existed on Mars, various experiments were conducted at the end of the twentieth century to replicate a Martian environment and test its potential to support life. Now the closest terrestrial equivalent to Martian soil is volcanic ash. So experiments were conducted to see if 'methanogen' bacteria can survive in ash from a Hawaiian volcano. Timothy Kral and his colleague, Curtis Bakkum, grew four species of methanogens in conditions resembling the environment three kilometres beneath the Martian surface (Coghlan, 1999a: 14). It might be speculated that if methanogens can survive on Mars they could be employed in a terraforming project whereby these bacteria would release methane to create a greenhouse effect, warming the planet and releasing its frozen water.

A rather eccentric suggestion for the greening of Mars is to bomb Mars with Trident missiles containing CFC gases. Objectors to this proposal draw attention to technological problems, one of which is that launching the number of rockets, large enough to carry a sufficient load to 'green' Mars, would have an adverse effect on the Earth's atmosphere. There is also the objection that some of the suggested methods of terraforming Mars might dry up water reserves quicker that the present natural processes have done.

In the course of terraforming Mars many problems would have to be overcome, but no laws of nature would have to be breached and it would not require any as yet unheard of developments in technology. However, the resourcing plans for colonization are daunting. Governments and large investors have great problems in resourcing large-scale projects lasting more than one generation, as can be seen with reference to a ten-year project like the Channel Tunnel, which has obvious potential benefits.

Venus

Venus is roughly the same size as Earth but different in three important respects: first, its atmosphere is 100 times thicker, and is composed primarily of CO_2, with a high surface temperature due to the greenhouse effect. CO_2 is more efficient in letting heat in than letting it escape. Second, Venus does not have global plate tectonics like Earth, which means that its interior heat source produces more catastrophic volcanic activity. Third, there is no surface water; all the Venusian water is in its atmosphere.

The task of terraforming Venus is very daunting. The first objective would be to produce a more clement temperature. With sufficient technical skills it might be possible to move asteroids into orbits that would result in collisions, where each impact would release some of the atmosphere from the planet. The problem with this suggestion is that even if there were a sufficient number of available asteroids the impact would devastate large amounts of the planet's surface.

Other proposals for the reduction of the Venusian temperature include the use of a gigantic sunshade to reduce solar energy, or to add dust particles to the upper atmosphere to absorb or reflect heat from the Sun. Even more exotic are proposals to send genetically modified microbes that can convert CO_2 into carbohydrates. But even if all of these proposals were feasible, the terraforming of Venus would take thousands of years, which means that large-scale financial support for such a venture would be unlikely.

Other locations?

There are proposals for terraforming Io, the innermost of the Galilean moons of Jupiter. According to James Oberg (1995: 86–91), it has several advantages over other candidates. Foremost is its source of internal heat generation caused by tidal stresses induced by Jupiter. However, among the drawbacks is the deadly radiation belt which surrounds Jupiter. Also, Io has no water and no atmosphere, but the radiation belt could be decontaminated by pulverizing one or more of the smaller moons, says Oberg, thus creating a ring of rocky debris which would provide protection from Jupiter's radiation. Water could be imported from Io's icy companions, Ganymede and Callisto. According to Oberg (1995: 89), Io could be habitable by the end of the next century.

Ethical aspects of terraforming

At present terraforming projects are unfeasible, but in the long term they might fall within the scope of possible programmes. In which case, the ethical issues they raise will require elaboration. The ethical arguments for and against terraforming other planets can be summarized as follows.

In favour of terraforming

1 A terraformed planet would provide a base for astronauts, scientists and colonists; a possible new frontier or refuge from catastrophic events on Earth, such as natural calamities or war. Mars could support human settlements. Although smaller than Earth, it has much land space. A new frontier might generate new world values, such as co-operation, and provide a point for criticism of old world values. From a social perspective there is a chance

that 'new world values' may develop among the early settlers who would apply social and political ideas that could be exported back to the home planet.

2   Terraforming is a long-term project for humans, generating scientific advances and inevitable beneficial spin-offs in technology. But the greatest benefit would be the knowledge derived from attempts to control the planet's climate. Even if, for the first few hundred years, the project was no larger than one of the Antarctic bases, the knowledge derived from attempts at terraforming could be applied to projects to protect the Earth's ecology.

3   Life has an intrinsic value and consequently it is morally commendable to permit life to flourish on planets were previously there has been none.

4   Massive terraforming projects might divert scientific and technological resources away from warfare and destructive objectives, kindle enthusiasm for international cooperation, and restore humankind's lost confidence in its ability to create.

Against terraforming

1   The arguments against terraforming other planets are based on moral objections: is it morally right to cause such drastic changes to another planet? Mars might contain as yet undiscovered forms of microscopic life that have evolved under Martian conditions, and consequently our presence will jeopardize these by bringing our micro-organisms which may attack and kill the Martian ones. This objection would be stronger if there was evidence of Martian life-forms, as a major climatic change could also upset their development.

2   There is also an objection that human efforts to transform Earth have resulted in a catalogue of man-made disasters and unforeseen catastrophes. How much worse would it be if we started in an environment of which we know less than we do of Earth? Something might go wrong, leaving things even worse off with regard to the planet's ability to foster life. There might even be repercussions on Earth.

3   The fact that terraforming is a long-term project would act as a disincentive to governments with regard to investment. Moreover, scarce human talent and resources would be diverted from worthy projects on Earth, such as social and environmental problems.

4   If terraforming and hence colonization are successful, they would not divert resources away from warfare: on the contrary, wars would very likely be fought over the new territory; and military uses of the new colonies would simply extend the arena for socio-political problems.

While most of the above arguments for and against terraforming are likely to remain unresolved for some time, there is a strong case for saying that terraforming projects should be postponed until it is certain that there are no native forms of life that might be harmed in the process. This problem is central to any form of exploration, whether it is the Antarctic or the planets. Hence international treaties are essential to ensure that Mars and other sites are not contaminated by terrestrial biota during exploration. In fact, this problem was recognized in the Viking surveys when the two spaceships were assembled in controlled clean rooms and then subjected to an alcohol wipe-down of external surfaces and other components, to remove biological material, and finally heated to sterilize their interiors. In the event this proved unnecessary as the intense UV radiation on Mars would have killed off any bacteria.

One thing is certain: as demographic pressure increases on Earth, and as its natural resources are used up, proposals for planetary colonization will be offered and considered. Maybe the technology which has devastated large parts of Earth will continue to exploit and devastate the planets. Or maybe that very technology will put Mars into reverse and restore it as an abode for life. The technology used to colonize Mars could one day be employed to save the inhabitants of Earth from an ecological catastrophe.

Beyond the solar system?

A Martian settlement could, because the planet's gravitational field is weaker than Earth's, provide a suitable launch for further space exploration. But journeys to the stars would require craft operated on principles of physics we have not yet conceived of, including exotic assaults on time and space. Return trips to the stars would be prohibitive in one human life-span, and even with the predicted time-dilation effects of travelling near the speed of light, a traveller would return after a few years only to find out that thousands of years had passed on Earth. This could very well prove a disincentive to such travel.

To appreciate the problems, however, it is important to reconsider the vast distances and time involved in interstellar travel. With current rocket systems a return trip to Alpha Centauri would take over 200,000 years, and it would probably take much longer to find a habitable planet similar to Earth. It is obvious that stellar colonization will involve thousands of generations who will live their entire lives engaged in space travel. How would they survive and reproduce in conditions approaching zero gravity? Sexual relations in these conditions might sound like a novel activity, but there is too little known at present on the effects of near zero gravity on a human embryo. Apparently human reproductive systems encounter problems in space: altered light and dark cycles inhibit male testosterone and sex drive; and the female menstrual cycle can become erratic or cease altogether (McInnis, 1999: 42–3). Perhaps artificial insemination supported by chemical agents would resolve these problems, but the predicted deterioration of muscles and organs during prolonged periods of weightlessness might result in future generations who are incapable of surviving in gravitation similar to Earth's, should they ever reach an Earth-type location. Meanwhile, on the journey, problems related to inbreeding, risks of epidemics and so on, would have to be solved. There is no reason why these problems cannot be solved, and maybe they will be overcome and human beings will survive intact, ready for the physical hazards of frontier life on a new planet. Or maybe, having left Earth behind, future humans will have adapted so well to their space environment that they will prefer to remain in nomadic colonies in permanent space flight.

Proposals for large-scale rockets where people will spend several generations before arrival at their destination raise serious ethical problems. Future unborn generations will be condemned to a restricted life with all the hazards and uncertainties of interstellar travel. But if we consider in-flight colonies which retain a system of conscious communication with Earth, or were completely self-contained, which succeeding generations would see as home, the above objections could be dispelled. Moreover, the voyagers might have no alternative once it is known that their sun is about to become a red giant. It might then be seen that space colonization is not like European colonization of the nineteenth century, where a large degree of cultural continuity survived the journey. Instead, it might be speculated that the eventual colonization of space, in which the colonization of Mars is the first step, is more like the movement of organic life from the sea to land. This would require a revision of our ideas of colonization and a maturity of outlook which was not attained by the Europeans of the nineteenth and twentieth centuries.

Rockets or microwaves?

There is a view which holds that space exploration and colonization are unrealistic and that we should discount analogies with colonial expansion during Earth's history, but rather concentrate our efforts on electronic communication and data exchange. Why build rockets and engage in high risk human expeditions when we can explore the galaxy with probes and electronic equipment? The era of space travel began at the height of the Cold War when rival political powers sought to demonstrate their superiority. Some thirty years after the first Moon landing, the main justification for space exploration was still in terms of terrestrial political objectives, such as a demonstration of US technological superiority over the former Soviet Union and the ultimate domination of the 'free market'. Is what Kennedy and others saw as the first step in the conquest of space to be the last great scientific technological achievement? Is it time to put an end to notions of conquest and colonization inspired by terrestrial political rivalry?

The opportunity to explore outer space may soon pass. Some estimates suggest that we only have about thirty years left before a combination of factors, including space debris, electromagnetic pollution and energy limitations prevent further attempts at space exploration (Gott, 1993). Space exploration is expensive and projects for manned expeditions to far-off planets are likely to be curtailed in a cost-conscious society. Nevertheless, it is sometimes argued that expenditure on space travel could be a good alternative to the arms race during the current lull in hostilities between the superpowers. There is certainly a need for public involvement in large-scale non-aggressive projects. There are, of course, other contenders for these resources. Massive levels of poverty, hunger and disease, dogmatic fundamentalism and ethnic rivalries, threaten world peace and place burdens on human skills and resources. Yet proposals for the redirection of resources to much needed human welfare projects are, at present, an invitation to commit political suicide. Space explorations could present a challenge to a culture which is tired and lacking a sense of direction. But the tiredness of our culture – evidenced by a toleration of bankrupt political authority and economic theory – may only be one explanation of society's unwillingness to resource large-scale projects for manned exploration of space.

It may be that we are seeing a decline of the nineteenth-century belief that travel and great expeditions add to the wealth of human knowledge. Scientific legends, such as Darwin and Wallace, travelled to the colonies in search of scientific truth while artists and writers, such as Gauguin and Conrad, sought inspiration in travel. Perhaps, as we enter the twenty-first century, the link between travel and knowledge will finally be severed. Why did scientists and artists of the nineteenth century travel and mount expeditions to remote parts of the Earth? One answer is to obtain information. But once the necessary infrastructure has been installed, information can be transported anywhere, at little cost, at the speed of light. One does not need to appeal to science fiction to predict that the scope and nature of information technology will contribute to a transformation of our attitudes to knowledge throughout the twenty-first century. There are many who proclaim that the simulation of experience by cyberspace technology could eliminate many reasons for being physically present at a particular place or event. For example, with a telecommunication presence on Mars supplied by a robot probe, covering various regions of the planet, it should be possible to provide a virtual reality experience of Mars to homes and schools all over the Earth without any of the attendant risks associated with costly space voyages. Is this the future of space colonization? Dominance in information technology, it might be argued, will replace competition for the old-style notion of colonial dominance which required the physical presence of the ruling civilization.

The promise of the information technology revolution has placed limits on the notion of space travel. Notwithstanding its ring of satellites, the information technology culture is primarily Earthbound. In fact, the proliferation of satellites required by information technology may place limits on the feasibility of space travel. NASA has predicted that if the amount of space debris exceeds 150,000 fragments of one centimetre or larger, it would render space flight impossible. Each satellite launch increases space debris, and to achieve high speed access to the Internet will require several hundred more satellites (Ward, 1997: 49). Moreover, the growth of information technology has also removed one of the incentives to travel; we have no need to travel in order to know. Some have argued that it is time to shift our models of learning. Edward Purcell provides an interesting corrective to our view about travel and experience.

> Suppose you took a child into an art museum and he wanted to feel the pictures – you would say, 'That isn't what we do, we stand back and look at the pictures and try to understand them. We can learn more about them that way'.
>
> (Purcell, 1980: 196)

This could be a useful way of redirecting our ideas about space travel. Scientists who are currently involved in the search for extraterrestrial intelligence are convinced that intelligent beings will explore the universe, not with spacecraft but by means of data exchange.

There are, of course, limits to Purcell's analogy; the child or art student may come to appreciate the paintings by standing back and admiring them, but there is no substitute experience other than actually being in the presence of great works of art. Knowledge of the great works of art requires direct confrontation: photographs, films and forms of electronic data exchange are very inferior media for aesthetic experience.

It might be argued, however, that our need for direct confrontation is merely because existing levels of technology provide inadequate representations, but with a vastly improved technology, direct experience would not be required. For example, probes could be sent out to space, bearing a very advanced form of an electronic equivalent to our senses, and the information so derived could be directly relayed to us. Would, for example, this 'virtual' experience of another planet be fundamentally different to an actual visit to the planet in question? Could human exploration be replaced by robot space probes, which would in effect be an extension of our senses? The objection is that many major technical breakthroughs would be required before these virtual simulations could be perfected. This raises philosophical questions concerning the amount of perfection that is required to transform virtuality into actuality. Is there such a thing as a perfect VR sim; that is, a perfect simulation of our five senses working together, or is it a conceptual impossibility? This is not the kind of question that can be resolved with reference to telescopes and microscopes, where technology greatly enhanced our perceptual experience, but it is a question concerning a major reappraisal of our concepts of experience, sensation and knowledge.

Before we allow ourselves to be carried too far on the promise of information technology we should remember that the inhabitants of Earth face wars and civil unrest as well as shortages of material resources, and the potential destruction of the Earth's ecological balance. These problems will not be solved by increasing our capacity to transfer information across the galaxy. While there are many proposals to avoid the potential environmental catastrophes – recycling and environmental protection schemes – there is still the dream of escape from a tired and self-destructive world. Many people still yearn for a return to nineteenth-century expansionism and colonization – maybe next time we might not repeat the mistakes of our forefathers but expand with humility and respect for the universe.

Notorious setbacks, such as the loss of contact with the Mars Observer satellite in 1993; the loss of the Mars Climate Orbiter in 1998 and the Mars Polar Lander in 1999, were high profile disappointments, revealing huge system and management errors, including the catastrophic fouling up of English and metric units in the case of the Mars Climate Orbiter. Critics have also blamed NASA's 'faster, better, cheaper' policy which may herald a withdrawal from some ambitious space projects. Meanwhile, space science has no alternative in an era of cost-containment government control and the absence of competitive stimulation after the Cold War. But failures will not deter space exploration; and it has been argued that the faster, better, cheaper policy allows for a quicker response to failures and successes and that the overall Mars programme is sufficiently robust to cope with failures and adapt to changes. The Mars Surveyor 2001 Lander has been cancelled and NASA is re-examining plans for the 2003 Lander and Sample Return. Missions planned for 2005 and beyond are likely to suffer delays but it is hoped that NASA will learn from past disappointments. Failures will not deter space exploration. The hunt for evidence of former life on Mars has given a significant impetus to exploration and, during the next few years, numerous attempts to explore Mars and other solar planets are inevitable.

In many respects it is lack of funding rather than scientific imagination which prohibits further developments in space exploration. There are advanced plans for Moon bases, manned flights to Mars, permanently occupied space stations and colonies. There is no shortage of proposals, most of which do not require unheard of scientific breakthroughs. Although the costs of space exploration are massive, the current expenditure is relatively small. The annual NASA budget, for example, is less than two weeks of US defence expenditure.

With present technology it is obviously less expensive to explore the galaxy with microwaves, than with spaceships. But maybe these options are not mutually exclusive. Projects are being considered for human settlements in space and possibly on the Moon and Mars, and more economic methods of space flight are under discussion. Nevertheless, space scientists are very much their own worst enemy. Space exploration has been intimately linked to systems for delivering atomic weapons; and BION II proposals to return to the practice of sending primates into space have triggered widespread protests which have damaged the image of another generation of space scientists. There were widespread protests after a monkey, named 'Mutlik', died on 8 January 1997 in Moscow after returning from a 14-day space mission. It died following surgery to remove electrodes that had been implanted in its body before flight. The bad habit of torturing animals for minimal scientific gain is hard to break and costly projects facing well-organized public dissatisfaction may not survive.

The penetration of space, its eventual colonization, will be an essential feature of what Michael G. Zey (1994) has described as the coming 'Macro-industrial era'. As evidence Zey cited plans for big engineering projects by the USA, Japan and Europe which will include a Moon base early in the twenty-first century followed by a Mars colony by the year 2019. Current proposals are less ambitious; it is more likely that the US Martian missions in the early decades of the twenty-first century will involve robots rather than humans. However, if it transpires, the Macroindustrial era will be characterized by a return to programmes of economic growth with massive wealth creating engineering projects including the exploration and colonization of space, thus liberating humanity from the home planet. This would involve terraforming projects on potentially habitable planets, like Mars, and the eventual transformation of other spheres close to the Earth. A large-scale commitment to space exploration and colonization would, argues Zey, offer an alternative to an information technology-dominated culture which has swamped the planet with white-collar bureaucracies and paper, where annual reports become quarterly and then daily, and on-line information and junk mail pervade everywhere. In the early days of the 'computer revolution' there were wild predictions of a 'paperless society', as computers were destined to replace paper in schools and offices. The reality turned out to be paper nightmare. It is certainly time to reconsider the promise of the information technology revolution, which has been accompanied by a decrease in overall prosperity fuelled by the mistaken belief in the obsolescence of such once great industries as aircraft, ship-building, steel production and textiles. These industries, and projects arising out of them, once created wealth, but there is little evidence that information technology can enrich the lives of more than a small minority, and even less evidence that its fruits will improve our standard of living, bring peace or solve any of the major problems facing humanity.

There is no disputing the fact that information technology will play a tremendous role in the transformation of human culture, but it raises important philosophical questions concerning the nature of freedom and truth. What might be called the Internet ideology involves a belief in freedom from physical entrapment into a cyber dimension. But those who inhabit this dimension forget that freedom is bound up with the physical bonds which tie humans together and tie us to nature, our history and the institutions and physical structures that have been created throughout history. Perhaps the time has come to re-think objectives and look beyond the information culture together with its nihilistic post-modernist philosophy. If this is the case, then large-scale space projects could be undertaken and not dismissed as outdated objectives and prohibitively expensive.

Of course, this kind of thinking runs counter to the current politico-economic wisdom which appeals for zero growth or sustainable growth – two synonyms for stagnation and austerity. Yet the gains from space exploration could be significant: a return to massive wealth-creating projects, with technological spin-offs which already include the manufacture of light-weight materials now used in wheelchair construction; communication technology from space satellite technology; pharmaceutical research from space laboratories, including zero-gravity laboratories in which it is possible to grow protein crystals for the development of new drugs that may one day combat AIDS and other diseases. Because of the initial investment costs, space projects will require international co-operation, employing skills and resources from many cultures, thus possibly contributing to a global spirit of peaceful co-operation.

Opponents of space science are usually dismissive of appeals to its spin-offs or serendipitous results. Frequently cited in this context are velcro, teflon and the fruit drink, Tang. It is pointed out that these gains might well have been achieved if sufficient resources had been directed at the problem, rather than indirectly through chance. Does the non-stick frying pan justify the billions spent on Moon rockets? Why not channel resources directly into improvements in the manufacture of cooking utensils? Why spend billions on space programmes when there are urgent problems to be tackled here on Earth? According to Gonzalo Munévar (1998) this kind of objection fails to appreciate the necessary role of serendipity in the evolution of science. Although serendipitous results cannot be guaranteed, they are part of the essentially transformative nature of science. For scientific inquiry is not merely a fact-gathering, problem-solving process, where data are collected in order to solve our immediate problems. Rather, it is part of the process by which our present position is likely to be transformed. Thus new scientific developments play an important role in determining the range of problems and opportunities we may encounter. But these are not always predictable. For example, Munévar points out how laser technology has dramatically changed medicine, giving opportunities for new surgical techniques and therapeutic intervention. But the revolution within physics which led to the discovery and use of lasers would never have come from medicine, no matter how much support could have been given to surgeons and medical researchers. Munévar consequently insists that serendipity is an inevitable feature of science and that space exploration, as part of the development and transformation of scientific ideas, will inevitably provide both tools and ideas that can 'change our panorama of problems and opportunities' (ibid.: 179). He suggests that new developments in astronomy – radio astronomy, superior forms of Earth-based telescopes, X-ray and Gamma-ray observatories – could be 'particularly helpful in pointing to areas of physics where new directives will be fruitful' (ibid.: 189).

According to many of the arguments developed in this book, space travel will be essentially linked to the search for extraterrestrial signs of life. Just as the presentation of claims concerning the existence of microfossils in Martian meteorites led to a wave of proposals for manned missions to Mars, so would evidence of an intelligent signal from space lead to intensive attempts to visit the source.

5

UFOS AND SPACE TRAVEL

> The phenomenon of UFOs is real and we should approach it seriously
> and study it.
>
> (Mikhail Gorbachev, during a meeting with workers in the Urals,
> reported in *Soviet Youth*; Lebedev, 1991: 64)

Introduction

Orthodox scientific opinion, including that of SETI researchers, totally rejects
accounts of visits to Earth by extraterrestrials. Few scientists are willing to bestow
credibility on theories which purport to explain how extraterrestrials could visit
us, and the branch of investigation known as Ufology is branded as pseudo-
scientific nonsense. Yet belief in alien visits, in flying saucers, contacts and
abductions, has a vast public following, and discussion of UFOs is, perhaps, the
public's first introduction to the subject of extraterrestrial life. This chapter
examines purported evidence of UFO visits and various reports of encounters
with extraterrestrials.

Reports of UFO experiences

From earliest times there were beliefs that the heavens were populated, usually by
Gods who travelled in supernatural chariots or by spiritual assistance. Only after
the Second World War was it finally recognized that rockets provided the most
likely means of travel within and beyond the solar system. But this had a
sobering effect on those with aspirations towards space travel. The limitations
imposed by energy requirements produced more modest predictions of interplanetary exploration. The most popular conception of an alien spacecraft is the
'flying saucer', so named by Kenneth Arnold, a 32-year-old civilian pilot who
reported some over the Cascade Mountains on 24 June 1947. This report made
newspaper headlines. Within a week flying saucers were reported in every US
state, as well as in Canada, England, Australia and Iran. There were thousands
of subsequent reports and researchers were later to discover many alleged

sightings which pre-dated Arnold's report such as the reports of 'Foo fighters' – luminous balls that followed aircraft – by Second World War pilots. Shortly after Arnold's sighting, a military aircraft crashed during a search for an alleged UFO. The debate surrounding this crash generated massive interest and fuelled beliefs in a military cover-up with regard to information about ET visitors. Most reports of alien spacecraft are based on a fleeting sight of objects in the sky. Less numerous are accounts of landings, contacts and occasional abductions, where witnesses usually report being treated as medical patients.

Several years ago a Gallup poll in the UK revealed that one in five educated people believe in alien UFOs, while a similar survey in the USA revealed a one-in-two ratio of believers and a 1973 Gallup poll revealed that 11 per cent of the American public claimed to have seen a UFO (Randles and Fuller, 1990: 12). In 1986 a survey by the Public Opinion Laboratory at Northern Illinois University found that 43 per cent agreed with the following statement: 'It is likely that some of the unidentified flying objects that have been reported are really vehicles from other civilizations' (McDonough, 1987: 184). A US poll of MENSA members, reported in the *National Enquirer* on 1 June 1976, showed that two-thirds of their members believed that UFOs are 'spaceships from another planet'. A report in the *Observer* of 13 August 1995 cited an annual rate of 2,000–3,000 UFO reports to the Ministry of Defence in the UK and suggested that the real amount of 'sightings' may be higher as many people fear ridicule or don't know how to file a report (Sweeney, 1995). Yet scientific orthodoxy remains highly sceptical of Extra Terrestrial Activity (ETA). After over fifty years of intense interest, no witness has produced an authentic photograph, alien artefact or piece of testable knowledge from such a source. In fact, one of the problems with UFO reports is not merely whether the information is true or false but whether all of the information has been given. Dozens of people claim to have been aboard extraterrestrial spacecraft but none have brought back any extraterrestrial tools, artefacts or any knowledge that could not have been derived from terrestrial sources.

Do these reports convey real experiences? The answer to this question depends on what is meant by 'real' and 'experience'. For many witnesses they are indeed very real experiences. A helpful distinction in this context can be drawn between 'inward experiences' and 'outward experiences'. The former can be examined by reference to the state of mind of the participant; the latter, if real, would require corroboration by independent sources. Reports of alien spacecraft have been discussed in almost every developed nation on Earth. From a scientific viewpoint this fact itself is an extraordinary social phenomenon. The psychologist C.G. Jung proposed that alien spacecraft reports represented a strange manifestation of the human psyche. Others have argued that they are in some way associated with an external agency.

Many investigators prefer to speak of Unidentified Flying Objects (UFOs) rather than flying saucers. During the past fifty years they have been reported all over the world with widespread publicity. This publicity has rendered it almost impossible to obtain observational reports that have not been influenced by previous sensationalist, and invariably explanatory, reports. However, when strict observational criteria are employed, few cases of UFOs turn out to be genuinely unidentifiable. On investigation some 95 per cent are eventually classified as identified flying objects (IFOs) and the argument turns on the status of the remaining 5 per cent.

The American astronomer, J. Allen Hynek, a former sceptical UFO investigator who was for twenty years a scientific consultant to the US Air Force inquiry known as Project Blue Book, became convinced that some reports require further scientific investigation. Hynek was Professor of Astronomy at North-western University and Head of the Center for UFO studies at Illinois where, until he died in 1986, he had compiled some 100,000 entries from over 140 countries in his data bank. His definition of a UFO was:

> the reported perception of an object or light seen in the sky or upon land the appearance, trajectory, and general dynamic and luminescent behaviour of which do not suggest a logical, conventional explanation and which is not only mystifying to the original percipients but remains unidentified after close scrutiny of all available evidence by persons who are technically capable of making a commonsense identification, if one is possible.
>
> (Hynek, 1972: 10)

This definition, claims Hynek, should enable investigators to distinguish between misperceptions of known objects, like weather balloons and kites, and those which require more intensive scientific investigation. It calls for investigation and evidential support within the framework of current scientific theory and technology. It does not make any reference to evidence or observational methods which might be relevant from a standpoint associated with a different disciplinary matrix, and it does not, as T. Patrick Rardin (1982: 257) points out: 'make any distinction between evidence that is available in fact and that which is available in principle'. Without extending the meaning of 'in principle' too far, so as to include the manifestly implausible and esoteric, the Kuhnian point can be made that new theories and explanations may initiate new evidential requirements. This is precisely the problem with Ufology, as the residue of anomalous observations do not fit easily with current theories of motion and transport. Here an appeal to Kuhn's model of incommensurable paradigms can be made where different explanatory frameworks are 'not only incompatible but often actually incommensurable with what has gone before' (Kuhn, 1962: 103).

The essence of the philosophical problem concerning evidence for UFO experiences is whether they can be dealt with by the methods appropriate to normal science, or extensions of it, or by means of a new explanatory scheme or paradigm in Kuhn's sense. There are certainly problems with many traditional forms of presenting evidence. Eye witness reports, photographic evidence and films can be challenged and are prey to accusations of forgery. Photographic evidence is, by itself, of little value. Photographic forgeries have been common since the beginnings of photography. Recent techniques of digital imaging, where a picture can be reduced to a series of pixels (digitized picture elements) and be entirely manipulated by a computer, raise fundamental questions about the value of photographic evidence. Conventional photography has always been open to manipulation and forensic scrutiny of photographs purporting to represent encounters with aliens has usually revealed evidence of interference. But a digital image can be manipulated without detection and it is not beyond the range of modern computer-assisted photography to produce a fairly coherent photograph of a UFO on the White House lawn with the President greeting its inhabitants. Artefacts and wreckage of alien spacecraft would count as hard evidence, but these are either unavailable or their alleged existence is a matter of dispute.

The very concept of 'hard evidence' is problematic. UFO sceptics insist on hard evidence but in doing so forget that successful disciplines such as astronomy do not rely upon the hard evidence provided by singular convincing cases, but by the steady accumulation of explanations, theory and evidence.

Evidence of close encounters

Initially, Ufologists classified close encounters (CEs) with extraterrestrial biological entities (EBEs) in terms of four kinds, but Stephen M. Greer (1992) added a fifth. They are:

First kind	CE1: an observation of a UFO at a distance less than 500 feet.
Second kind	CE2: where some trace of observable evidence is obtained, such as an artefact, piece of wreckage or landing marks on the ground.
Third kind	CE3: observation of an EBE, generally within the vicinity of a UFO.
Fourth kind	CE4: interaction with an EBE where the observer is taken aboard the UFO, sightseeing visits and abductions.
Fifth kind	CE5: intentional interactive communications, a response to a human initiated signal, such as flashing lights in reply to human signals.

Evidence of CE1s is derived from large numbers of sightings from numerous 'reliable' witnesses, together with frequent sightings of photographic evidence and ground tracings. The small town of Bonnybridge in Scotland, which has a population of 5,500 people, produced 2,000 UFO sightings in three years. These are usually descriptions of bright flashing lights, and while the majority have been explained, there is still a residual 250 unexplained and the local MP has supported calls for a public inquiry (Mullin, 1995:2–3). Evidence of sustained government interest is frequently cited in support of the claim that 'there must be something in these reports'. Yet observations have frequently been explained in terms of natural phenomena or a human origin. Witnesses have been shown to be unreliable. The photographs and other evidence could have been of other sources than spacecraft, for example natural phenomena or hoaxes. So far there is no evidence that governments have any knowledge about UFOs other than rumour.

Evidence for CE2s is also problematic and controversial. There are on record at least twenty-eight reports of UFO crashes between 1942 and 1978 (Bord and Bord, 1992: 69). They are usually reported to have taken place in deserts and remote jungles, such as the Sahara Desert, the Bolivian jungle and uninhabited parts of New Mexico and Arizona. Reports of military cover-ups frequently accompany these alleged 'crashes'. One often cited example came from Roswell, New Mexico. On 2 July 1947 there was a report from an isolated ranch, some thirty miles from the nearest town, concerning a crashed UFO. Later reports suggested that four alien bodies had been recovered near the wrecks. Witnesses and various reports claimed that alien bodies were taken to Area 51, an unacknowledged test site in Nevada. An early press release said that a flying disc had been recovered, but this was later denied in an official statement which said that it was two weather balloons. But military interest and a veil of secrecy encouraged the belief that it was an alien spacecraft. It was also argued that the military were concerned to maintain a high security cordon because the object was an experimental launching of a German V-2 rocket that had gone astray.

The case for public disclosure of the Roswell incident was eventually taken up by the New Mexico Congressman, Steven Schiff. In response the Air Force produced a report in 1994, which acknowledged that the wreckage did not come from a weather balloon, and that this had been an earlier cover-up story. In fact, said the report, it was a top secret Pentagon balloon designed to detect sound waves produced by Soviet nuclear explosions. In a report in the *New Scientist* Richard Weaver, Director of Security at the Air Force, said: 'This research indicated absolutely no evidence of any kind that a spaceship crashed near Roswell or that any alien occupants were recovered therefrom' (cited by Pearce, 1994: 4). The same article also quoted Walter Haut of the UFO museum near Roswell, who saw the Air Force Report as a 'straight continuation of the cover up' and maintained that the alien technology is so advanced that government officials do not want to admit that it exists. Still unsatisfied with the Air Force's report, Congressman Steven Schiff requested a search of government files relating to the incident by the General Accounting Office (GAO). Schiff believed that the Air Force may have concealed information. In August 1995, the GAO reported that while they saw nothing to contradict the Air Force's earlier examination, they did discover that some government reports from that time have been destroyed (Kiernan, 1995: 10).

In 1996 Channel 4 TV broadcast a film of a post-mortem performed upon four alleged alien bodies recovered from the Roswell wreckage. In the same programme witnesses claimed to have identified four aliens at the scene of the crash. The film of the post-mortem was allegedly shot by a US cameraman. This film appeared in London in 1995, but to date no meeting has been held between Channel 4 TV and the elusive cameraman. Moreover, no sample of film is available for laboratory testing which could determine the year of exposure and processing. In July 1997, on the 50th anniversary of the incident, the Pentagon published the results of a four-year multi-million dollar investigation into the Roswell incident, *Roswell Case Closed*, which says that the wreckage was of a high altitude balloon with four life-sized dummies which simulated parachute drops. This was part of the US Army's 'Operation Mogul', devised to detect Soviet nuclear tests in the upper atmosphere.

None of the accounts so far offered to explain the Roswell incident have produced enough evidence to rule out rival explanations. While the Pentagon insists that it has quashed all rumours, a poll conducted in 1997 indicated that 65 per cent of Americans believe that a UFO crashed outside the town of Roswell. The Army press officer who originally issued the release on 8 July 1942, later told ABC news that: 'any dummy knows what a dummy looks like and those weren't dummies' (*The Times*, 25 June 1997: 15). So controversy and speculation continue. Nevertheless, if an alien spacecraft had crashed and this had been followed by a government 'cover-up', then it is likely that someone would have leaked the story by now. But then – would anyone believe them? Leaks do not convince sceptics, who can simply dismiss them as disinformation programmes. At the heart of the Roswell case is a gut feeling of scepticism exhibited by American people towards the pronouncements of their government and the scientists who serve it.

One piece of evidence of alien visitors which might be relevant is the discovery of a skull in the Commonwealth of Independent States (CIS) in 1991, of which Sh. B. Begaliev, Head Physician at the Panfilovski Regional Hospital said, 'belonged to a creature with a highly developed intellect, but of non-human origin' (Lebedev, 1991: 70). This opinion was based on 'the large volume of the cerebral hemispheres, and the fact that there were no normal eye-sockets or nasal holes', and a low volume of calcium in the bone structure. (ibid.: 70). Like many others of this kind, reports of further and more comprehensive investigations do not appear in UFO literature.

Reports of CE3s and CE4s are numerous throughout the world and an annual compilation has been provided by Timothy Good (1991, 1992, 1993a). But much of the evidence is frustrating, despite the fact that many reports have more than one witness. Witnesses are frequently held to have been momentarily paralysed by the EBEs, or their recording equipment damaged, such that hard evidence is unavailable. There also appear to be significant cultural variations in these reports, suggesting the influence of terrestrial concerns. For example, during the 1950s and 1960s interactions were reported to be friendly, but during the past twenty years reports of more hostile encounters have increased, with abductions of a very violent nature. This may say more about our cultural criteria than about EBEs. Scientific knowledge may also have played a role in the changing experiences of CE3s and CE4s. During the 1950s and 1960s EBEs described their origins within the solar system, but as knowledge of the solar system grew, and the likelihood of life on nearby planets decreased, then EBEs were reported from further afield and many reports are obscure regarding their origins. Although there are many abduction reports, almost all have a strong imaginary element.

Reports of abduction by EBEs are numerous and one investigator, Jenny Randles (1988), described the problem as involving near epidemic proportions from people of many different backgrounds. Susan Blackmore (1994a) refers to a survey which reports that 4 million Americans have claimed to have experienced an alien abduction. A report in the *Observer* (30 June 1996: 15) stated that up to 150,000 Americans are insured against abduction by extraterrestrials. However, the author noted that it is not clear as to how someone could make a claim if it happened or whether a large number of claims would push up the premiums. One frequently cited abduction involves the US Ufologist, Linda Napolitano, with two security guards and a 'third man' – alleged to be the former UN Secretary General, Javier Perez de Cuellar. His excellency denies the abduction (Sweeney, 1995: 25). But are these abduction reports something to be explained or explained away? Are they real or a new form of psychosis? If real, then to what kind of reality do these abduction reports refer? What kind of experience is an abduction experience?

To date, the physical evidence supporting abduction reports is inadequate. Wounds and scars in themselves offer no proof that they were caused by EBEs. Abductees often report that EBEs have placed implants in their bodies, but these are either expelled or lost. So far, examinations of stains on the clothing of abductees have not proven to be of extraterrestrial origins, and no extraterrestrial artefacts have been identified.

Early abduction or CE4 accounts appeared in the 1950s in the books by George Adamski, *Flying Saucers Have Landed* and *Inside the Spaceships*, where the author recounted his visits to Venus and the far side of the Moon where he saw rivers and plants. Adamski's stories, and many that followed, bear little resemblance to modern CE4 accounts and abduction experiences. Many abductees recall their experiences under hypnosis, where accounts of medical examination are revealed. It should be noted that hypnosis is not a guarantee of truthful reporting, as such recollection need not be 'real' lived in the world experiences, but rather recollections of mental images.

Appeals to the phenomenon of 'false memory syndrome' challenge the validity of evidence derived from interviews conducted under hypnosis as well as those conducted with conscious abductees. Studies have shown (Blackmore, 1994b) how the imagination can help to reconstruct memories. Sleep paralysis has also been proposed as a possible explanation, where the dream mechanisms operate while the body is paralysed. The problem with this explanation is that many abductees report experiences that occurred when they were awake. Susan Blackmore (1994b) investigated research by Michael Persinger, a neuroscientist at Laurentian University of Sudbury, Ontario, who claims that abduction experiences, along with other psychic experiences such as out of body experiences (OBEs), are linked in some way to excessive bursts of electrical activity in the temporal lobes. According to this explanation, people with a 'high temporal lobe lability', prone to unstable temporal lobes and frequent bursts of electrical activity which can be seen on an EEG, fall into the category of abductees. At present this explanation does not have sufficient power to rule out other explanations of abduction experiences. As Blackmore notes, Persinger's observations 'are nothing more than correlations'. Moreover, those who believe that there have been genuine abductions point out that many abductees do not score high on measures of temporal lobe lability. But at present, the main objection is that not enough tests have been carried out to support either Persinger or those who offer other explanations.

Nevertheless, Persinger has sought an element of predictability in his attempts to simulate abduction experiences. Blackmore describes his experiments involving the application of magnetic fields across the brain which have produced strong emotions of anger, fear, and a sense of being physically manipulated. Were a similar experience felt outside of a laboratory it may well be possible, says Blackmore, for it to be 'remembered' as an abduction, especially if the abductee was 'cued' by a hypnotist or interviewer.

Randles (1988) outlines several common features in modern abduction reports. First, there is the phenomenon of missing time, where a sense of time is lost as in a fantasy state, or some form of sensory deprivation experience. Randles describes this as the 'Oz factor', after the experiences depicted in the film *The Wizard of Oz*. It is, however, a notorious feature of abduction reports that many are associated with long night-time driving, which amounts to a form of sensory deprivation. Second, it has been noted that under hypnosis many 'witnesses' recall childhood CE4 experiences. Third, many reports echo familiar SF stories and films, but do so selectively. There are no reported encounters with Mr Spock, the Klingons and other popular aliens. Fourth, many of the 'witnesses' have good relations with publishers, although many have been destroyed by the attendant publicity and ridicule.

Randles discerns a pattern to the abduction reports. They begin with the capture, then move on to the examination, and then various exchanges which may often include a message, and then the experience ends. Abduction accounts, she argues, unlike myths and folk stories, do not follow national stereotypes. The description range of EBEs is narrow: they are usually three and a half feet tall, with big round or pear-shaped heads, round eyes and slit noses and mouths. They can also be six feet tall, thin, Nordic, beautiful and blond. The small ones are unfriendly, although not frequently hostile; whereas the tall are magical, philosophical, and enjoy communicating messages (Randles, 1988: 181–2). According to Randles, abduction experiences are some kind of 'objectified dreams' – real physical events – involving 'a system of symbolic imagery which suddenly erupts into the three dimensional world' (ibid.: 220). She also offers the intriguing suggestion that maybe aliens are creating the experiences, not by coming here and abducting people, but by using the minds of the witnesses, using methods to control consciousness over vast distances. Clearly, there is a need for more studies of abduction reports and CE4s. But as in much of Ufology, the presence of hoaxers and fraudsters is a fundamental barrier to any serious study.

Encounters of the fifth kind (CE5) are considered feasible by the scientific community, particularly among SETI workers who attempt communication by means of radio signals, although it should be stressed that no SETI scientist has yet claimed to have established contact. Other claims regarding CE5s have emerged from devotees of the paranormal. In recent years the phenomenon of channelling – a modern equivalent of spirit messages – has been associated with EBEs. Channellers claim contact with The Elohim, a group of spacemen, who are recorded in the Book of Genesis. The Elohim have been allegedly contacting terrestrials since 1973, when they decided that we needed to be taken under control. Apparently they have no problem with communication as they speak all the languages of the world. Generally, the messages received offer various brands of leadership and salvation to a world that is already suffering from a surplus of political tyrants and despots. Another group of channellers receive messages from 'The Council of Nine' who, since the early 1970s, have been represented by a messenger called Tom (Schlemmer and Jenkins, 1993). Other channellers receive messages from the Intergalactic Federation and the Ashtar Command. The latter occupy a fleet of spacecraft which have been in orbit for millennia. The messages are notorious for their banality and unoriginality, with appeals to save the world from nuclear war or pollution, appeals to a higher form of spirituality or consciousness, which is usually linked to the requirement that the contactees or terrestrial recipients of the messages are treated as unique individuals. The channellers offer the standard message associated with religious cults: the world is threatened and the chosen few will be airlifted to safety at the right moment. All of this is wrapped up in platitudes about higher levels of awareness, living at one with the Earth and various other cosmic forces. If these messages are really forms of CE5, then contact with EBEs is disappointing.

Many claims regarding extraterrestrial visitors rely heavily upon techniques of persuasion, some of which ought to be separated from claims appertaining to evidence or theory. There are frequent appeals to pioneers like Galileo, whose rivals refused to look through his telescope and consequently failed to appreciate the momentous discoveries we now attribute to him. Just as Galileo was ahead of his time and unappreciated by his narrow-minded contemporaries, so it is argued, today's unappreciated Ufologists are tomorrow's Galileos.

There is, of course, an abundance of historical evidence of the sheer wrong-headedness of experts who remained sceptical of new developments. George Stephenson, the locomotive engineer, was ridiculed by his detractors. In fact, the engineer, Thomas Tredgold, wrote in 1835 that the possibility of 'any general system of conveying passengers ... at a velocity exceeding 10 miles an hour, or thereabouts, is extremely improbable'. On these terms, it might be argued, many claims with regard to new theories of transportation which transcend current knowledge should be seriously considered.

This argument misses the reality of the history of science. Most of yesterday's crackpots remain crackpots. Galileo and those like him succeeded, not because their theories were provocative but because they supplied answers to questions, addressed the issues of the day, and their theories proved increasingly successful and applicable.

Another method of persuasion involves the linking together of known mysteries and resolving them all with one grand solution. The assassination of President John F. Kennedy and the alleged cover-up after the sudden death of Pope John-Paul I have been linked to theories concerning extraterrestrial visitors (Andrews, 1987).

The appeal to prestigious witnesses and celebrities features largely in UFO literature. It is a fact that most societies place greater value on some categories of witness over others. Ufologists place greater emphasis on observational reports from military and civilian pilots – now estimated to exceed 3,000 – as well as the police. There are good reasons for this, as the members of these professions are frequently highly trained in observational procedures. Independent multiple sightings from such sources should consequently be taken seriously. What does not deserve serious consideration are the numerous reports of observations by celebrities, some of whom may have career interests in the association with topical events. Among the more dubious appeals to celebrity observations is one made by George C. Andrews, author of *Extra-Terrestrials Among Us* (1987), who cites British newspaper reports that Prince Charles had observed a UFO near Windsor Castle on 9 March 1986. The fact that the Prince did not deny the report after it was published is also seen by Andrews as further confirmation of the sighting, although he does not mention that the Royal Family once had a tradition of not responding to media statements about them. It is worth noting that Windsor Castle is on the flight path to Heathrow airport – one of the world's busiest air spaces!

During the 1960s and 1970s several books by Erich von Däniken (1969, 1970, 1973, 1974) developed the theory that Earth was visited in early times by astronauts from another solar system. The ancient astronaut theory claimed that many of the deities worshipped by ancient people were in fact ET visitors. The argument usually relied on claims that many impressive buildings, such as the Great Pyramid at Gizeh, stone relics of the Incas and the Mayas, or Stonehenge, were constructed under the influence of ET visitors. In the same vein, ancient writings of biblical prophets and cave drawings were cited as evidence of encounters with extraterrestrials. In the absence of evidence demonstrating that no technological civilization with the capacity for space travel has ever existed in the history of the universe, it cannot be proved that astronauts have not visited

Earth in the distant past. Consequently, von Däniken's theory would seem to be possible and probable. There could be a strong case in favour of conducting a scientific or historical inquiry to test the ancient astronaut hypothesis. This would involve an examination of claims that certain constructions could not be replicated by human beings. But having established that humans could not have achieved any of these feats without assistance would not establish a case supporting the involvement of astronauts. It still leaves open the possibility that fairies, ghosts or any of the recognized deities in the world, could have lent a hand. The ancient astronaut hypothesis does not have sufficient strength to rule out other explanations.

Von Däniken's opponents responded by demonstrating the force of Occam's razor, suggesting that ordinary hypotheses should be considered before invoking extraordinary explanations. Thus archaeologists and historians have demonstrated that early societies were capable of manufacturing the artefacts which had been attributed to ET visitors; it was shown how thousands of slaves could cut and drag the stones for the Great Pyramid and erect them in place. Similar explanations have been produced with regard to the construction of Stonehenge and other relics. Ancient texts and early art works have also been explained in terms of representations of natural phenomena, rather than the extra-human sources associated with ET visitors.

Few, if any, SETI researchers subscribe to any of the ancient astronaut theories. Most of von Däniken's accounts of ancient ET visitors have been repudiated by historians, theologians and archaeologists. No data that can be used in a novel way has been discovered. Not one claim about ancient ETs has passed a simple test which requires the production of a piece of knowledge that was unavailable at the time of the visit. All that is needed is just one statement about the speed of light in an ancient tomb, a part of a computer or a laser. Most of the examples cited by von Däniken, his references to ancient drawings, or writings, are open to several interpretations. Despite his appeal to consider the ancient astronaut hypothesis as a basis for investigation, none of von Däniken's hypotheses have the potential to rule out rival hypotheses, which is what is sought in either a good historical or scientific explanation. It is possible that astronauts visited the Earth, but this hypothesis is only one of a possible infinite number of explanations of the phenomena it attempts to explain.

Aside from the methodological problems with ancient astronaut hypotheses, they mean that the rich evocative mythology of early religious texts, symbols and truly amazing constructions, are reduced to banal science fiction, with spacemen, rocket launches and landing sites, together with deities armed with ray-guns and atomic disintegrators. Even the great mysterious edifices, like Stonehenge or the Pyramids, are reduced to points in an extraterrestrial energy distribution system akin to the national grid.

One of the superficial strengths of theories regarding extraterrestrial visitors is that of proving a negative; we can never *prove* that they and their UFOs do not exist, any more than we can prove that ghosts and leprechauns do not exist. But even when claims of UFO sightings are demonstrably mistaken, one encounters the fact that evasions of falsification frequently abound in UFO literature. But it is not just the evasion of falsifiability that causes problems; it is the appeal to the fantastic, the levels of paranoia, on which the *ad hoc* evasions rest. Lack of evidence is often concealed with reference to government and military cover-ups and disinformation plots. Here the imagination can run riot. Andrews (1987: 159) suggests that George Adamski, whose notorious blend of fact and fictionalized nonsense discredited Ufology, was actually a CIA disinformation agent. There are reports that astronauts have encountered UFOs and the fact that these reports are later denied by the astronauts in question is cited as evidence of a cover-up. Following many reports of contact, key witnesses are frequently alleged to have been silenced after visits from men in black suits. These particular visitors appear after a UFO report and either threaten witnesses against further revelations or remove evidence of the encounter. Sometimes the men in black suits are said to be government agents whose objective is to maintain the veil of secrecy over alien encounters, but on occasion they are held to be the EBEs themselves. Claims supporting the latter view are supported by witnesses who note the puzzlement displayed by the men in black suits to ordinary household items, their mechanical speech and other forms of alienness.

Problems with 'nuts and bolts' ET craft

One of the strengths in the sceptic's position is the fact that the burden of proof must fall upon exponents of hypotheses concerning extraterrestrial visitors. For the massive obstacles in the way of this form of transport are obvious to the scientifically literate. The first obstacle concerns the sheer difficulty in imagining how a craft can travel such large distances. The enormous distances involved make reports of frequent visitations hard to comprehend. A light year is the distance light travels in one year of 31.56 million seconds at the rate of 186,000 miles per second, which is a distance of 5.88 million miles. The nearest known star, Proxima Centauri, is 4.2 light years away. The fastest rocket with current technology would travel at 25,000 mph, and would take 300,000 years to reach Proxima Centauri. Most theories of travel are inhibited by these distances. No vehicle, however propelled, can travel faster than the speed of light, according to Einstein's theory which, until recently has been widely accepted by the scientific community. Even if the speed is scaled down to 10 per cent of the speed of light, visits to the stars would be prohibitive: a journey to Sirius would take eighty-eight years, which would be a one-way trip for terrestrials. However, intelligent EBEs may have solved the problem by circumventing the limitations of Einstein's theory, or developing techniques of suspended animation, or hibernation, or by simply living longer lives.

At present no one has any idea of how to manufacture a craft that could approach anything like the speed of light. Even if this speed were attainable, a voyage across the galaxy would last for decades and the energy required would be beyond any foreseeable technical ability to produce it. According to Einstein, the faster the object moves the greater will be the increase of its mass. At a speed approaching that of light a vehicle's mass would increase tremendously requiring near impossible levels of energy to propel it. Moreover, at speeds approaching that of light, collision with microscopic particles of stellar dust would be catastrophic and the craft would require a massive shield which would also add to its payload. Nevertheless, it has been argued that faster-than-light (FTL) travel is actually compatible with Einstein's general theory (Crawford, 1995), although there are no practical solutions to the problem of how to manufacture and fuel such a craft.

Proposals for FTL travel often involve appeals to 'warps' and 'wormholes'. These concepts deserve consideration. Warps or 'warp-drive' sound like the stuff of science fiction although they are permitted by the laws of physics. Apparently warp-drive involves the manipulation of space-time such that it expands behind the spacecraft and contracts in front of it so that a spacecraft travelling fairly slowly will 'push' its departure time back and 'draw' its arrival time forward. Now this might be permitted by the laws of physics, but its practical application would require something approaching the energy output of a black hole.

Travelling through wormholes is another proposal for FTL travel. In 1915 Ludwig Flamm saw hints of the possible existence of wormholes in Einstein's equations. Wormholes can be explained as follows: consider a black hole so massive that its powerful gravitational field curves space to such an extent that within it there is a singularity – where the curvature of space-time is so infinitely sharp that the laws of physics break down. Now imagine two singularities which are somehow linked together; for example, one at point A and another at point B, which is thousands of light years away. Presumably, if they are linked, then if one entered the mouth of the wormhole at point A one could instantly emerge from point B. Apparently wormholes are not disallowed by the laws of physics, although this does not suggest that they actually exist. The problems with wormhole travel are obvious. If they involve faster-than-light travel, there is the objection that it would allow causality violations, such that time travellers might go back and prevent their own birth. So it has been assumed that some law of physics prevents their actual occurrence. This now leaves unresolved the question whether FTL travel is permissible without the causality violations.

Entering a black hole is rather dangerous. Nevertheless, the laws of physics do not prohibit the possibility of joining a black hole with a white hole, one that works on opposite principles to the black hole and expels matter. But then the problem is that a white hole is likely to be unstable and decay rapidly, making it very improbable for a spaceship to emerge. Generally, black holes are regarded as dangerous and unstable places, but some theorists maintain that they can be kept open and stable if they are crammed with 'exotic matter' which, however, can be created out of nothing when space and time are curved the right way (Matthews, 2000: 12). So far, speculation regarding wormholes and exotic matter lacks observational support, and is justified only insofar as they are 'permitted' by the laws of physics. Perhaps the future lies in the actual discovery of a wormhole or the possible manufacture of wormholes in a laboratory.

Marcus Chown (1996) considers travel through wormholes and acknowledges that while Einstein's general theory of relativity allows for the existence of shortcuts through space and time, it is a far step to proving their existence and constructing them. However, Chown notes that two lines of inquiry are under consideration. The first is the attempt to construct a microwormhole in a laboratory; the second involves a search for wormholes that have either survived after the big bang or have been created by advanced extraterrestrials. The remarkable properties of wormholes would make them detectable from Earth, as light from stars would fluctuate as the wormhole mouth passes between the Earth and the star.

Conventional spacecraft, however, must meet tremendous problems of energy consumption. Even if it is assumed that extraterrestrials are more technologically advanced than us they will still have to cope with a near 100 per cent mass-energy conversion near the speed of light. To appreciate the problem, assume a level of 100 per cent energy conversion efficiency for a 1,000 tonne payload on a round trip from the nearest star to Earth at 70 per cent the speed of light. This would require energy equivalent to 500,000 years worth of the total annual electrical power produced by the USA (Tarter, 1990).

Of course the raw material for the craft's energy sources might not have to come from terrestrial sources. Gerard K. O'Neill (1975) suggests that a system of satellite solar power stations, manufactured in space and located in a geosynchronous orbit from Earth, would not only provide enough energy for Earth but provide enough energy for space flight with no loss of Earth's resources. These solar power stations could, he says, be initially manufactured out of lunar material. If this were possible for terrestrials, a similar solution to the energy problem could be devised by extraterrestrials.

NASA is currently developing forms of Solar Electric Propulsion (SEP) which could enable a wide range of missions that would be financially prohibitive with the standard chemical propulsion fuel available at present. Instead of burning a chemical propellant, the SEP system operates with a very high-velocity stream of ionized xenon which gently propels a spacecraft over a long interplanetary cruise. The advantage is that SEP requires much less propellant than conventional rockets, although it does take longer for its gentle thrust to build up high velocity. Consequently, it is more appropriate for lengthy voyages or colonies on the move.

If, however, we suspend restrictions on space travel imposed by current technological limits, it is possible to conceive of a craft which is coherent with current scientific theory but well beyond our present skills. There has been speculation about the feasibility of a photon craft, whose energy would be derived from the annihilation of matter by antimatter – an antiproton plus a positron. This would be the most efficient rocket fuel imaginable, as the mutual annihilation of matter and antimatter leads to a 100 per cent conversion of mass into energy. If enough could ever be produced and stored it would supply 10^{10} times the energy of the same weight as gasoline. Travelling at a velocity slightly less than the speed of light, it would take about ten years for a round trip to the Centauri system which is just over four light years away. This would require a vehicle weighing about 1,000 tonnes to provide the power and living facilities for a crew of twelve. The problem with this proposal, and many similar, is that it requires many unforeseeable breakthroughs. At present there are no ideas on how to obtain and store antimatter. Speculation on the latter merits a very low plausibility rating. The same can be said of proposals to build ships which scoop up matter and process it into fuel on the journey. This raises problems regarding the design of the scoop and the availability of matter.

One solution to the energy problem would be to travel slower with large spaceships designed like small island worlds sending out small probes. It is possible that extraterrestrials could approach the Earth in this manner and that their large ships could exist relatively undetected quite near to us. After all, Pluto is a very small planet in our solar system which has only recently been observed. Artificial worlds might well exist undetected on the outer edge of the solar system. Many SETI scientists, however, reject this argument. Frank Drake, who maintains that an interstellar quarantine exists with regard to space travel, argues that such a project would require millions of sophisticated probes, and then a wait for thousands of years until they functioned perfectly and were monitored by the home planet. In contrast, argues Drake (Drake and Sobel, 1993: 132), 'the Universe permits only limited kinds of direct encounters among its residents'. Thus he concludes:

> It makes most sense to me to ply the radio course – to plumb the radio spectrum for magic frequencies and scan the stars for beacons sent by alien intelligence. Information-laden radio messages are the quarry we seek in the search for extraterrestrial life.

Thus SETI researchers reject ideas of direct contact with spacecraft but seek deliberate signals or leakage radio signals. Whereas space travel is expensive and slow, electromagnetic radiation carries information at the speed of light at little cost. Among SETI scientists there are no expectations of contact by means of visiting spacecraft: it is information, not things, that will be exchanged between intelligent extraterrestrials and human beings.

There are certainly major physiological problems with the idea of human-like beings travelling on interstellar journeys. Lengthy journeys in prolonged states of weightlessness raise the problems of potential damage to heart muscles, as all under-used muscles shrink. Effects of zero gravity on humans can be debilitating. The Russian cosmonaut, Valery Polyakov, spent 438 days on the Mir space station in 1994 and from his experiences researchers learnt that astronauts who stay in space lose 1 per cent of their bone structure each month, suffer from a major deletion of muscle mass and face disruption of nerve centres. Longer-term stays would involve loss of blood cells and anaemia. Shannon Lucid, a US astronaut who spent 188 days on board the Mir station, required six months physiotherapy on return. Research has also shown that the stress involved in spaceflight actually weakens the immune system of the astronauts. Thus a virus that would normally remain dormant could, due to a weakening of the astronaut's immune system during a lengthy voyage, initiate a serious illness (Coghlan, 1999a: 15).

There would also be problems of previously undiagnosed malignant diseases, such as a brain tumour developing during a long trip and affecting a vital crew member. Accidents or illnesses requiring surgery would yield unique problems in conditions of weightlessness. For example, blood will not ooze or flow into the surrounding cavity to be mopped up, but will migrate from the wound, forming a red mist. Anaesthetic gases would put the surgeon to sleep, and an intravenous drip would not work. Psychological problems associated with long periods with the same individuals have been reported among Russian space crews, although better profiling techniques might avoid these kind of problems in the future. The problem of finding the right kind of crews is significant. The 'right stuff' sought in the early astronauts is clearly incompatible with lengthy space voyages. The early astronauts were test-pilots who required action and stimulation, but the long trips of interplanetary travel would require people who would not be prone to boredom. One solution to boredom might be the development of recreational drugs with no harmful side-effects for long-duration space flights. There are very serious psychological problems for humans who endure long periods of isolation. These could become critical when radio contact is cut off when the spaceship is the opposite side of the Sun from the Earth. Nevertheless, there is nothing in the human mind or body which could not be dealt with in order to make journeys of several years possible. Of course beings with a non-humanoid physiology and psychology might not be affected by these factors.

Even if colonies were established within the solar system, the energy requirements for lengthy space flights would still be massive. In this respect the appeal to large numbers of sightings actually weakens the argument for visits from EBEs. For even if they have established a colony on the Moon – the nearest possible site – it would still require a massive energy output to maintain a rate of visits in proportion to even a small fraction of recorded sightings. This also raises the question, why are they manufacturing so many craft? Why are they coming here so frequently and behaving in such a pointless fashion? And why, after at least forty years of visits, having mastered levels of technology beyond us, have they not made their intentions comprehensible? The answers to these questions remain in the realm of speculation. Ufologists differ over the reasons for UFO visits. Some have argued that their purpose is benevolent and that they merely wish to observe us and collect information. It is sometimes suggested that they may be concerned over our warlike ways and are ready to intercept if we look likely to wipe out intelligent life. This can be regarded as the 'Saviour hypothesis', which has certain superficial resemblances to western religious beliefs.

Another answer is that the UFOs are actually von Neumann probes (Tipler, 1980) and that the vast number of sightings which, if true, would indicate the use of large amounts of fuel and construction material, which could, perhaps, be explained with reference to Tipler's postulation of self-replicating probes using local resources (see Chapter 6). If this is the case, there is no evidence that they are using terrestrial resources; there are no alien mines, factories or assembly plants.

Theories of galactic colonization are popular with science fiction writers but the practical problems are clearly enormous. Consider spaceships travelling one-tenth the speed of light which are committed to a programme of galactic colonization. Suppose they spread outwards at distances of 50 light years, and then allow about 500 years recuperation time to consolidate settlements, extract resources and establish outposts for further exploration. Each 50 light years of colonization would take 1,000 years. The diameter of the galaxy is 100,000 light years, which would require 2,000 stages. At 1,000 years each it would take about 2 million years to colonize the galaxy. Of course, the galaxy has been in existence for considerably longer than that, and several scientists have speculated on the possibility that a colonization process is already underway. This topic, however, will be examined in Chapter 6.

Explanations for overcoming the distance and energy requirements vary in plausibility and credibility. Gary Kinder's (1988) report on the experience of the Swiss farmer, Eduard Meier, offers one explanation. Throughout the 1970s Meier claimed to have had frequent contact with beings from a distant cluster of several thousands of stars, known as the Pleiades, in the constellation of Taurus, which is about 500 light years from the Earth. Astronomers maintain that the Pleiades are too young for intelligent life to have evolved. But Meier responds by stating that they have formed a colony there and are adapting to the planetary climates artificially. Even if there is intelligent life there the distance is impressive. Travelling at the speed of light a return trip would take 1,000 years. According to Meier's reported 'conversations' with the Pleiadian visitors, it is possible for them to travel many times faster than light and on average they could make the trip in seven hours by using an advanced technique of collapsing time and space (Kinder, 1988: 185). Further details of this 'advanced technique' are not supplied, which is why such reports are of little value for scientific explanation. But Meier's credibility sinks with his claims to have travelled backwards in time, to have photographed the great earthquake at San Francisco, to have met Jesus and to have been inducted as the thirteenth disciple, and to have taken a trip to the edge of the universe where he photographed the eye of God (Bord and Bord, 1992: 170).

It is, of course, always possible to dismiss objections to long-distance UFO journeys by means of appeals to systems and principles 'not yet appreciated by terrestrial scientists'. These frequently refer to gravity or anti-gravity technology, hyperdimensional energy sources, techniques involving time dilation, travelling through 'wormholes', and various ways of taking a short-cut through space and time, as well as consciousness-assisted technology (CAT) or technology-assisted consciousness (TAC). As it is believed that many advanced EBEs have mastered telepathy, precognition and other senses which are beyond our knowledge, they can presumably utilize the energy of consciousness.

Parallels are sometimes drawn between scientific sceptics who dismiss UFO reports by citing energy limitations on interstellar travel and nineteenth-century scientists whose prejudices prevented them from investigating reports of meteors, despite well-documented evidence. Antoine Laurent Lavoisier had pointed out that stones could not fall from the sky as there were no stones in the sky! Meteors were observed, and the craters were there to be examined, but scientific sceptics ignored this evidence because they believed that rocks cannot fall from the sky. So, argue the Ufologists, today's prejudices regarding the energy limitations on interstellar journeys, will be swept away by the discoveries of future generations of scientists. Moreover, they may argue, such advanced technology may be at the disposal of the EBEs whose visits are so well documented.

The problem with this argument is that a shift in scientific beliefs to the acceptance of meteors required very little revision of theory – no modifications to classical mechanics – and did not require any prior commitment to beliefs which were wholly incompatible with canonical knowledge. This is not the case with appeals to as-yet-unknown energy sources. Moreover, there is no prejudice against the discovery of new energy sources as there was against claims concerning meteors; the world is crying out for new energy sources and would amply reward any scientist who discovered them. But sadly, there are no such sources and no plausible suggestions as to how they could be found.

The main objection to UFO space flight is bound up with arguments about the restrictions on terrestrial organized space travel: it is too costly in resources, and solutions to energy problems require exotic new scientific breakthroughs, although plausible developments in nuclear fusion engineering capabilities could be anticipated within a century or two (Kuiper and Morris, 1977). The other main objection is that the distance between potentially habitable sites is too great. Of course this might not be such a problem if they have life-spans much longer than ours or bigger spaceships possibly providing more living space than our crowded cities, so that several generations may dwell there on long voyages or travel in states of suspended animation for centuries.

Not every account of UFO phenomena is attached to beings from far-away places, who travel in 'nuts and bolts' spacecraft. There are theories of super-beings who live under the sea, advanced civilizations in remote parts of the world such as the Himalayas or the Antarctic, under the ice or even in parallel universes to which access is found in various hypothesized 'tunnels' (Bord and Bord, 1992: 186). One proposal by David Barclay (1994) actually invokes the traditional philosophical problem of the relationship between appearance and reality by suggesting that UFOs are actually 'telepresences in a universal VR'. According to Barclay, UFOs are part of a cyberspace system, and the whole universe is itself a virtual reality system which encompasses our reality as well as

UFOs within some super VR arcade. The problem with this hypothesis, as with several other accounts of UFOs as 'visitors' from other 'realities', is the daunting task of designing a procedure to verify it.

Ufologists embrace methodology

In the early days of Ufology, researchers appeared too eager to verify sightings, which they then interpreted as evidence of 'nuts and bolts' spacecraft piloted by intelligent EBEs. Like numerous deities and other extraterrestrial visitors, EBEs are generally held to be concerned about human conduct. This concern was widely reported in the spate of UFO sightings after the Second World War and the beginnings of the nuclear age. Sensationalist reports merging with Hollywood fantasy led to a distancing of orthodox science from Ufology. Explanations offered by Ufologists frequently ignored Occam's razor, which is a rule against multiplying entities or – in general terms – a rule which says don't involve extraordinary hypotheses until the ordinary ones have been eliminated. The apparent resistance to falsification also contributed to Ufology's lack of credibility. However, modern Ufologists, such as Jenny Randles and Paul Fuller of the British Unidentified Flying Object Research Association (BUFORA), are strict adherents to Popperian-inspired scientific methodology, enthusiastically seeking to falsify EBE explanations and providing explanations which are acceptable to orthodox scientific opinion. In this respect the modern Ufologist is a debunker rather than a myth-spinning believer. Explanations in terms of atmospheric phenomena, hallucinations or hoaxes are generally expected from BUFORA publications. Over the years the BUFORA standpoint has been vindicated. So much 'confirmatory' evidence has been demonstrably unreliable. Photographs, which were once considered as hard evidence, are now held to have zero credibility because of the likelihood of fakes. With the advent of sophisticated image-manipulation computers whose work is undetectable, photographs unsupported by other reliable confirmatory evidence are unacceptable. Eye witness reports are also problematic as they are frequently influenced by psychological and cultural factors.

In several publications Randles and Fuller (1990, 1991) reconstituted observational reports of UFOs as evidential support for the atmospheric vortex theories propounded by Terence Meaden (1989, 1990, 1991) to account for the crop pattern phenomenon. Drawing on some of the remaining unexplained cases mentioned in the Condon Report and elsewhere, they offer a new meaning to these observations. Among the numerous examples of UFO data that could be reconstituted in support of an atmospheric theory is a BUFORA film, *Fire in the Sky*, which records a pulsating orange ball of light that drifted across the sky and was observed and filmed by a building surveyor, Peter Day, in Buckinghamshire on 11 January 1973. Other independent witnesses saw the same phenomenon, which has never been satisfactorily explained. Randles and Fuller (1991: 96) cite this as evidence in support of Meaden's theory of descending atmospheric vortices, which was originally developed to explain formations and patterns in crop fields (an examination and criticism of various competing explanations of crop patterns is found in Lamb, 1994).

It is well recorded how over 95 per cent of UFOs turn out to be IFOs which eye witnesses at the time had not correctly identified. These are usually aeroplanes, weather balloons, birds, kites, space debris, hoaxes, vapour trails, stars, planets, satellites and lightning. The post-stealth bomber, Project Aurora, is currently cited as an explanation for many UFO sightings. Many of the unexplained residual 5 per cent are so because they have not been investigated. There might not have been enough information provided, they might have been reported too late, or not investigated because of a shortage of resources or lack of interest. There is also a well-documented phenomenon where eye witnesses colour their observational reports with accounts of structured objects. These are often influenced by cultural expectations. Thus many eye witnesses described 'saucer shapes' after the media picked up the expression 'flying saucer' in 1947. But what is consistent in the raw data are reports of bright lights or fire glows, which may then be reported as fiery chariots, spaceships or flying saucers in the context of cultural expectations. This raw data, argue Randles and Fuller, could very well be of atmospheric origin and thus the remaining unexplained 5 per cent of UFO phenomena could be explained. They also attach an atmospheric meaning to the two great 'shining shields' in the sky which Alexander the Great reported when he tried to cross the river Jaxartes in 329 BC.

Despite the efforts of the BUFORA Ufologists there is a general aversion within the scientific community to UFO claims and requests to investigate them. Ufology is not regarded as a genuine branch of scientific inquiry and in general SETI scientists distance themselves from UFO hypotheses. But Ufology's main problem is not its lack of scientific method or alleged resistance to falsification. If that were the case, many other branches of science would be equally shunned. One problem with Ufology is that it has been so contaminated by fraud, misidentifications and sheer lunacy. Few scientists are willing to share platforms with cranks who claim to have had meetings with little green men from Venus, and claim to have travelled at many times the speed of light, or take seriously the evidence from witnesses who have been led by amateur interrogators.

Another obstacle to the recognition of Ufology as a genuine branch of scientific inquiry is the absence of predictability in an area of puzzlement where correct predictions would carry enormous influence. Evidence, by itself, can never carry the argument. What would happen, for example, if evidence of an EBE encounter were obtained, say, a widely acclaimed authentic photograph, video film, piece of wreckage, a few artefacts and scores of witnesses? It would very likely be dismissed and those concerned would spend the rest of their lives fighting off charges of dishonesty. They would not only be rejected by the scientific community but fellow Ufologists, with their own rival theories under threat, would resist their explanations. The only chance for acceptance would be a number of well-managed correct predictions with adequate pre-publicity.

Official reports on UFOs

The US Air Force have investigated UFO reports under several headings. These include Project Sign, 1948–9; Project Grudge, 1950–3; Project Blue Book, 1953–69. Although officially denied, there are claims that a further investigation, known as Operation Majestic, took place.

Project Sign

In January 1948 the USAAF initiated an investigation of unidentified flying objects, which was called Project Sign. It involved the collection of information about UFO sightings with a view to discovering whether they presented a security threat. In late July 1948, the staff of Project Sign prepared an intelligence report called 'Estimate of the Situation'. This was classified as 'Restricted – not Top Secret'. Several unconfirmed reports at that time and later said that the 'Estimate' concluded that many UFOs were interplanetary vehicles. But the report was never published and Air Force Chief of Staff, General Hoyt S. Vandenberg, to whom it was addressed, refused to accept it for publication because it allegedly lacked proof to support its conclusions. The report was ordered to be destroyed, although a few clandestine copies are said to exist. Project Sign made its final report, which was classified 'Secret' in February 1949, but it was declassified in 1961. It is thirty-five pages long and concludes that:

> Future activity on this project should be carried on at the minimum level necessary to record, summarize and evaluate the data received on future reports and to complete the specialized investigations now in progress. When and if a sufficient number of incidents are solved to indicate that these sightings do not represent a threat to the security of the nation, the assignment of special project O status to the activity could be terminated.

Project Sign, it would appear, shifted in emphasis from an investigation of possible ETA to a security inquiry in the context of Cold War military activity.

Project Grudge

Project Sign continued to investigate UFO reports until 11 February 1949, but in 1950 it was reorganized and its new title, Project Grudge, indicated a change of policy, in which an altered attitude was shown towards UFOs and UFO reports were evaluated on the premise that the existence of alien spacecraft was unlikely. Reports of sightings offered to Project Grudge were not accepted at face value and treated with a degree of scepticism. Project Grudge issued one report which was classified 'Secret' but declassified on 1 August 1952. It was concerned with a detailed study of 244 sightings; 32 per cent were considered to be astronomical objects; 12 per cent weather balloons; 33 per cent hoaxes; and a residual 23 per cent 'unknown'. There was a degree of controversy over the residue; some said they had not been adequately investigated, while others thought that they had. Among the conclusions of Project Grudge's report was a remark that future inquiries be reduced in scope. 'There is no evidence,' said the report, 'that objects reported upon are the result of an advanced scientific foreign development.' The main thrust of the report appears to have been an investigation into whether UFO sightings were in any way connected with aggressive attitudes of the Soviet Union. For the most part UFO sightings were dismissed as a misinterpretation of conventional objects, hoaxes, individuals seeking publicity, psychopathological persons, a mild form of mass hysteria or war nerves. There was concern – in the Cold War context – that UFO reports could be attributed to a form of psychological warfare. Fears had been expressed by the military that the enemy would issue a surprise attack when information channels were clogged up by UFO reports. Critics, holding positions which favoured ETA explanations, argued that the subject had been under-researched.

Project Blue Book

On 27 October 1951 Project Grudge was reorganized, but still under control of the USAAF. It was now called Project Blue Book and given the task of collecting all data on UFOs. There never was a Blue Book, this was the name given to the project which contained 13,134 reports on its files (Hynek, 1977: 8). Project Blue Book issued a series of status reports, initially classified 'Secret' but declassified as of September 1960, although copies were not available until 1968. It concluded that UFOs did not pose a security threat. These investigations took place against the background of high profile newspaper reports of sightings and a stream of publications concerning ETA. Many of these publications made allegations of military cover-ups. The USAAF did little to allay speculation and conducted their investigation in an atmosphere of secrecy, which only added fuel to cover-up theories. Project Blue Book finally concluded that: 'there has been nothing in the way of evidence or other data to indicate that here unidentified sightings were extraterrestrial vehicles under intelligent control' (Blue Book Information Office, 1968). Generally, Project Blue Book cited weather balloons and other balloons used as sporting devices, which can reflect light and travel at speeds of over 200 mph in high-altitude jet windstreams. Other explanations were in terms of psychological phenomena, such as autokinesis, which can be observed when a pilot stares at a fixed light in an otherwise dark environment and suddenly experiences an illusion that the light has begun to move erratically. Burns from jet engines, bright stars, planets, comets, fireballs, meteors and other celestial bodies were also deemed responsible, as were searchlights, birds, kites, clouds, spurious radio indications, hoaxes, ice crystals and fireworks displays. There was also an 'Insufficient Category', which included details of sightings where essential details were lacking. Sometimes, this might include reports without corroboration, as in the case of a sighting by one person in New York. Others in this category would include a failure to provide details of size, location or time.

Hynek, who worked as a scientific consultant to Project Blue Book, later complained of a reluctance to engage in a full investigation of UFO reports. He spoke of a 'can't be, therefore it isn't' attitude which permeated official thinking in both Project Grudge and Project Blue Book which meant that they had broken the cardinal rule of scientific procedure – investigate the facts before attempting to theorize. Hynek reported on the reluctance of the USAAF to investigate UFO reports after they were satisfied that there was no military threat involved. Scientists involved with the project, however, displayed an ambivalent attitude which Hynek (1977: 24) described as the 'committee complex' whereby 'a scientist will confess in private to an interest in a subject which is controversial or not scientifically acceptable but generally will not stand up and be counted when in "committee"'.

Other reports

Although the bulk of UFO investigation was conducted by the USAAF, official reports were conducted in other countries. In Canada a report was conducted by the Department of National Defence, and was known as Project Second Storey. This was conducted between 1952 and 1953. A low key operation was conducted by the RAF in England, who assigned one man to work with the Ministry of Defence where all sighting reports were sent. Sweden and Denmark also conducted low key inquiries.

The Colorado Project: the Condon Report

In 1966 the University of Colorado undertook a study of UFOs and Dr Edward Condon, Professor of Physics and Fellow of the Joint Institute for Laboratory Astrophysics, was the scientific director. The inquiry lasted for eighteen months and was published in 1968. Since then it has been known as the Condon Report. Although heavily criticized by Ufologists, the Condon Report was emphatic in its rejection of claims that extraterrestrial spacecraft were visiting the Earth. Dr Edward Condon, author of the report, was dismissive in his summary:

> Nothing has come from the study of UFOs in the past twenty-one years that has added to scientific knowledge. Careful consideration of the record as it is available to us leads us to conclude that further study of UFOs probably cannot be justified in the expectation that science will be advanced thereby.
>
> (Condon, 1968: 1)

Condon also reflected the opinions of many practical no-nonsense scientists when he outlined criteria for the acceptance of hypotheses concerning extraterrestrial craft. According to Condon:

> The question of ETA (Extra Terrestrial Activity) would be settled in a few minutes if a flying saucer were to land on the lawn of a hotel where a convention of the American Physical Society was in progress, and its occupants were to emerge and present a paper to assembled physicists, revealing where they came from, and the technology of how their craft operates. Searching questions from the audience would follow.
>
> (ibid.: 26)

One might even doubt whether this evidence would satisfy sceptics, as there is no mention of the requirement for the aliens to submit their report to a refereed and reputable scientific journal. Not to mention the fact that the extraterrestrials had not been funded by a reputable research contracting agency. But what would be the outcome if such a craft landed on the lawn of a hotel that was hosting an annual Ufology convention?

The Condon Report is a massive document covering over 900 pages of inquiry. Many reported sightings are either rubbished or explained in terms of conventional phenomena, such as aircraft, weather balloons, and so on. One typical example is an event in January 1968, near Castle Rock, Colorado, when thirty people reported UFOs, including a spacecraft with flashing lights. Observers even claimed to have seen the occupants who were presumed to have come from outer space. Two days later it was reported that two schoolboys had launched a polythene hot-air balloon in the vicinity of the sighting. But that was not the end of the matter. Reports taken from the original newspaper account of alien visitors were circulated to UFO journals, some of whom published them without reference to the explanation which followed. When criticism was made, one reply was that no one could be sure that all of the sightings were of the balloon and that there might well have been a genuine UFO as well as the balloon. This kind of argument, which clearly misplaces the burden of proof, has discredited Ufology as a serious branch of scientific research.

Critics of the Condon Report insist that too much emphasis was based upon the obviously dubious claims, leaving many anomalous observations under-examined. Hynek (1977: 286) and other investigators, have drawn attention to the summary of the Condon Report, which was dismissive of UFO observations and saw no need for further investigation, and the report itself which 'could not identify about one-third of the cases it studied'. Condon's definition of a UFO has also been criticized as it did not include the possibility of a distinction between observations which were unidentified by experts and those which were unidentified by the lay public, thus opening the floodgates to dozens of spurious reports which a scientific committee should not waste its time on. Condon (1968) defined an unidentified object as 'the stimulus for a report made by one or more individuals as something seen in the sky ... which the observer could not identify as having an ordinary, natural origin'. Hynek (1977: 286) maintained that this definition led to 'an inordinate waste of time and money', inflating the report with spurious observations which were easily identified by experts. According to Hynek:

> The sky is full of things which many observers find puzzling: bright planets, meteors, advertising planes, twinkling stars, etc. While it is true that the 'u' in UFO means unidentified, we must always remember to ask 'unidentified' to whom? A bright planet such as Venus shining through a cloud cover which is sufficiently thick to blot out the rest of the stars may appear strange and mysterious to a given observer, but it would not be to an astronomer.
>
> (ibid.)

Condon's criteria for proof of extraterrestrial involvement in UFOs can also be criticized for their one-sidedness. The example, cited by Condon, whereby conviction would be secured if a UFO landed on the lawn outside a hotel in which eminent scientists were in conference, offers a misleading concept of proof. Condon sought a single convincing case for ETI, but while in many activities dramatic evidence secures conviction, in much of science, especially in astronomy, conviction is achieved by means of a slow accumulative process which increases with acceptable data and theory.

Critics of the Condon Report fall into two camps. First, there are various New Age theories which involve a range of beliefs in parallel universes and spiritual dimensions which are unrelated to 'nuts and bolts' flying craft, and consequently it is argued that Condon had limited the scope of the inquiry to preclude these explanations. Second, there are scientific thinkers who maintain that Condon had abandoned critical observation, or that explanations of UFOs require a different methodology of science. The problem for scientific thinkers, who face pressure and ridicule from their peers, is that of preventing their objections to the perceived shortcomings of the Condon Report from collapsing into support for various New Age beliefs. Consequently, the scientific community has distanced itself from the UFO controversy.

Despite its critics the Condon Report marks a watershed in the scientific study of UFOs. Since then there has been very little interest within the scientific community for further investigation. The report, however, did not diminish public enthusiasm for details of alleged encounters with UFOs.

Operation Majestic

The status of Operation Majestic is very much a matter of dispute. The controversy focuses upon an alleged report compiled as a briefing for Dwight D. Eisenhower, President-Elect, on 18 November 1952. It is alleged to report on the wreckage of an 'alien spacecraft' recovered 75 miles north-west of Roswell, New Mexico, in July 1947. Four alien bodies were allegedly recovered, having apparently been ejected from their vehicle two miles east of the wreckage site. The name of the operation allegedly set up to investigate the incident was Majestic 12, or MJ-12.

It is still in question whether the 'leaked' documents concerning Operation Majestic are genuine. The report was allegedly prepared by Admiral Hillen-koetter, Director of the CIA. One document describes the four bodies: 'although these creatures are human-like in appearance, the biological and evolutionary process responsible for their development has apparently been quite different from those observed or postulated in *Homo sapiens*' (Good, 1991: 139; Blum, 1990). If authentic, this document would suggest that government sources were understood to have believed that beings from another world have visited Earth. In 1990 the UFO sceptic, Philip Klass, claimed that the MJ-12 documents were a forgery. Apparently there had not been a special investigation into the allegedly crashed UFO. This, however, was resisted by several Ufologists who claimed further levels of CIA disinformation (see Blum, 1990). There were claims that the CIA produced its own forgeries of the originals which were designed to be detected, thereby maintaining the secrecy of talks that have been taking place between US officials and aliens since 1947. In 1995 an investigation of government files relating to the Roswell incident by the US General Accounting Office concluded that the document known as MJ-12 is a forgery.

Military interest in EBEs

There are believed to be many political and military benefits of contact. Sagan (1970) has advocated space exploration as a useful diversionary occupation for military personnel in the superpowers. UFO investigations were actually encouraged during the Cold War period, when there were fears that the other side might make first contact and enjoy the advantage of knowledge of military hardware from a superior intelligence. The main explanatory theories within the US military during the period when Projects Sign, Grudge and Blue Book were operative, divided between hoaxes, ETI and Russian secret weapons. The potential military application of UFO research is significant. The military apparatuses of the superpower nations have long been engaged in a massive information search, collecting data on dolphin intelligence, paranormal spoon-bending experiments, and UFO sightings, very often in the belief that something just might turn up in a brute force search. Information on UFO sightings, for example, would be one more category in which to file less familiar data, and could usefully eliminate categories of unaccounted-for enemy aircraft. A similar system of classification might be employed for alleged ET signals from outer space. The reasons for collecting this data need not be based on the weight of evidence regarding ET life-forms, but on broad expectations unrelated to any specific results. Blum (1990: 58) reports on the 'something-might-turn-up' mentality of US defence: 'In the early 1980s Army Intelligence was an inventive why not-give-it-a-try sort of organization', spending 'millions on para-psychological experiments'.

In the Ufology literature are numerous reports of military cover-ups and disinformation policies – all suggesting that governments know of EBEs but will not tell. The reasons for the alleged cover-ups usually include [1] fear of panic and social disruption; [2] military fears that rival powers would gain access to the advanced technology which the EBEs possess; [3] a general mistrust of EBE objectives; and [4] fear of ridicule were the reports of EBEs to become recognized as a hoax. Timothy Good, who compiles and publishes reports of EBE encounters, argues that they have made contact, they are here, but that government cover-ups extend to the use of 'black budgets' to conceal the cost of such investigations (Good, 1993a).

A bureaucratic-inspired objective to collect and classify all data for no particular purpose can easily co-exist with military secrecy and a systematic programme of disinformation together with attempts to rubbish all serious forms of investigative research into ET-based phenomena. The seemingly paranoid assertions of Ufologists and ET watchers concerning military disinformation could very well be true, without lending any support to the usual inferences of secret military knowledge.

Military secrecy in this area can be accounted for in terms of three options. First, there is a high-level conspiracy; they know something that we don't know and are anxious to keep it to themselves. This, however, is unlikely as contact with extraterrestrials would be of such major interest that it would be impossible to maintain secrecy for long. Second, although there is no conspiracy, usual government furtiveness and paternalism are responsible for the withholding of data. This might be because of fear about panic and ballyhoo if certain suggestive data were released. One line of reasoning here can be traced to the public response to Orson Welles' *War of the Worlds* broadcast. A third reason for secrecy is fear of embarrassment. Major political and scientific figures could be lampooned by political opponents. Consider what the gossip columnists could do to a Minister of Defence if it were discovered that he once chaired a committee for UFO investigation. Just as Ufologists cite eminent witnesses to give credibility to their theories, so sceptics could destroy opponents by associating them with Ufology.

6

IS ANYONE THERE?

> But the barriers of distance are crumbling; one day we shall meet our
> equals, or our masters, among the stars.
>
> (Arthur C. Clarke, Foreword to *2001: A Space Odyssey*)

Introduction

Critics of SETI research have mustered very powerful arguments against
pluralism. Their arguments stress our uniqueness, the contingency of intelligent
life, maintaining that the odds are stacked against the possibility of a similar
intelligence to our own occurring elsewhere. These objections are considered in
the first part of this chapter. Thereafter the discussion focuses on appeals to the
fact that so far no contact has been made. This appeal is sometimes described as
Fermi's Paradox, so named after the Italian physicist, Enrico Fermi, who
responded to arguments in favour of ETI with the throwaway retort 'where are
they?' While it may be the case that just one authenticated contact would put an
end to this dispute it is, meanwhile, important to SETI research that a resolution
to Fermi's Paradox is found.

A critical appraisal of SETI

One of the early critics of SETI was the evolutionary biologist, George Gaylord
Simpson (1964a), who appealed to the uniqueness of intelligent human life,
maintaining that those attracted to bioastronomy were apparently unaware of
the contingency of human life. Despite the millions of diverse forms of life,
Homo sapiens only represent a tiny fraction of the possible forms it can take,
argues Simpson. A different start and an existing species would be different. 'If
the causal chain had been different *Homo sapiens* would not exist' (Simpson,
1964a: 267). Evolution is not repeatable: 'No species or any larger group has
ever evolved, or can ever evolve twice' (ibid.). The process of selection 'involves
long chains of non repetitive circumstances' (ibid.). In addition to the adaptive

contingencies are random factors in mutation, such that the process of evolution involves a whole series of non-repeatable accidents. According to Simpson:

> The assumption that once life gets started anywhere, humanoids will eventually and inevitably appear is plainly false. The chance of duplicating man on any other planet is the same as the chance that the planet and its organisms have a historical identity in all essentials with that of the Earth through some billions of years. Let us grant the unsubstantiated claim of millions or billions of possible planetary abodes of life; the chances of such historical duplication are vanishingly small.
>
> (ibid.)

Gonzalo Munévar (1989) has raised questions concerning the assumption in SETI research that extraterrestrial intelligence must be linked to an advanced technology similar to ours. He argues that, while one cannot reject the possibility that an advanced technology could evolve elsewhere, there is nothing necessary about the assumed connection between intelligence and technology. With a different natural history, an intelligence might have evolved without present technology, which relies heavily on human characteristics such as the kind of dexterity afforded by the development of matching thumbs and forefingers among other features. Furthermore, he argues, a technological culture is not an inevitable feature of human evolution. If other cultures had achieved ascendancy, then science and technology would not have emerged as a reference point for measuring intelligence. SETI's requirement for an almost identical technology, although scientifically understandable, is based on an impoverished concept of intelligence.

According to Munévar, the development of a scientific culture, with access to radio communication, is highly contingent, requiring a number of lucky breaks from the environment and human natural and social history. In this context Munévar cites the development of mammalian intelligence. It is widely believed that the dinosaurs were wiped out by the immediate effects of an asteroid or cometary impact or possibly volcanic eruption. But mammals who survived the years of darkness caused by the dust of the impact or eruption then evolved to occupy the niche held by the dinosaurs. If the dinosaurs had survived it seems unlikely that they would have gone on to develop a technological intelligence as they lacked the dexterity necessary for an elementary tool-making phase. But if the dinosaurs had not become extinct as the result of a major catastrophe, then mammals would probably have remained small vermin. In order to develop the size of their brains to facilitate eventual human intelligence they would require larger bodies. But long before their brains could develop enough to outwit the dinosaurs their large bodies would make them an easy prey. The emergence of the ascendant intelligence of humans was contingent upon the extinction of their predators.

Human ascendancy, Munévar argues, allows us to apply a brake on the development of high intelligence among other creatures. If racoons become increasingly intelligent they will be seen as pests and hunted to extermination (ibid.: 3). Our way of life prevents the emergence of high intelligence in animals that compete with us. Once the first species passed the threshold of tool- and weapon-making it pre-empted the possibility of other species evolving in ways that would compete with its supremacy. There are, of course, limits to our ability to control the development of our competitors. Despite our advanced technology we have not found a way of exterminating rats and mice and combating many forms of viruses and bacteria.

Scientific progress, argues Munévar, requires having the right physical environment. Our scientific breakthrough occurred when Newton unified terrestrial and celestial physics, bringing together astronomy and physics, making it possible for us to expect similarities throughout the universe. 'But in a planet very similar to ours but perennially covered by clouds,' argues Munévar, 'a comparable development of astronomy would be unlikely' (ibid.: 4). Electricity is essential to radio communication and a technological intelligence. In one form or another it has always been part of human experience, in lightning or electrostatic phenomena, inspiring people to consider ways of harnessing its energy. But on a planet with an extremely salty atmosphere it would not be observable.

The emergence of a scientific culture, Munévar points out, also requires a supportive social history. In our case science only just made it through the vacuum presented by the collapse of religion as a secular force. Galileo and his followers might not have pulled it off. A strong fundamentalist revival might well put an end to science. If the politicians phase out the arms race, there is a very strong likelihood that within twenty years progress in physics will have ground to a halt; and if they dramatically increase the production of weapons capable of global destruction, there is an equal likelihood that science will be terminated along with the rest of humanity.

Even if we grant the inevitable development of a technological civilization with access to Maxwell's electro-magnetic laws and the ability to manufacture radio transmitters, it is by no means certain that they will mean the same on other worlds as they do here. Munévar cautions 'against confusing an overlap in performance with an overlap in content' (ibid.: 6). Two radically different views may be concealed by the apparent ability to do the same thing. Despite a flat contradiction between ancient and modern astronomy regarding the position of the Earth, for practical purposes a sailor or navigator looking at the stars might well prefer either view. Both ancient and modern views allow an overlap in performance despite fundamental incompatibility of content. Likewise, if we receive radio transmissions from another species we should not conclude that these beings have the same understanding of Maxwell's laws that we do. There is, Munévar concludes, 'No inevitable, no highly provable connection between the appearance of life and that of intelligence, nor between the appearance of intelligence and that of an advanced technological civilization' (ibid.: 7).

Other critics have raised epistemological objections to SETI, arguing that its methodology is not as sure-footed as it sounds. Anthony Weston (1988) argues that SETI's reliance upon radio astronomy combines speculation with anthropocentrism; that it draws selectively from the evidence we have on how intelligence developed on Earth; that radio communication might not be as obvious a means of communication for ETs as it is for us; that magic frequencies, such as the hydrogen frequency, might not be their obvious choice, and that mathematically representable signals, such as a series of prime numbers, are not the most obvious candidates for an ET message.

It is, of course, conceivable that intelligent beings might not share our conceptions of logic and mathematics. John Taylor (1974) speculates on the existence of intelligent beings of only molecular proportions in a world which would have to be experienced in a radically different way than we experience it. For example, due to quantum mechanical effects it would be a probabilistic reality for them. Our logic, with its law of excluded middle – either it is or it is not – which is clear-cut, would be meaningless to them, as they would experience reality as a wide range of possibilities. They would never experience single occurrences, but only probabilities. Moreover, they would never produce a science or technology like ours. They could never produce a Newton, unless they grew to a size with which they could be affected by gravity.

According to Weston, when SETI scientists draw analogies between human intelligence and ETI they frequently ignore the past history of human 'contact' with other humans. Europeans discovered new lands which they perceived to be unpopulated, whereas in fact they were populated by civilizations of considerable standing. Not having experience with horses, the Incas saw mounted Europeans as one animal, akin to the centaurs of ancient Greece. Each generation of anthropologists accuses its predecessors of ethnocentrism. The history of human encounters with intelligent life, concludes Weston, is characterized by misperception.

Assuming that ETs have developed radio communication, how, asks Weston, can we separate their messages from other transmissions? The search must, nevertheless, be capable of distinguishing between intentional messages from advanced civilizations and leakage radiation from emergent civilizations. Consider the stars in Project Phoenix's 'target search', some of which are about 80 light years away. About twenty years from now they will be tuning into our leakage radiation consisting of broadcasts of the Goons and the first episodes of *I Love Lucy*, commercial TV, and eventually *Neighbours*, *EastEnders* and the fall of Margaret Thatcher. What conclusions will they draw from this? How will they separate reality from fiction? What kind of conclusions could we draw if the situation were reversed?

Probing SETI's assumption of an affinity between terrestrial and extraterrestrial intelligence, Weston raises questions concerning the preference for magic frequencies, like the hydrogen frequency or the Waterhole. Why should they share our ideas about what is general and simple? Although radio is a universal medium it might be limited to human culture. And even if ETs use radio, they might not wish to use it to communicate with us and might employ a code which prevents us from understanding their own internal communication.

Like Munévar, Weston stresses the contingency of Earth's scientific culture. Radio astronomy, he says, is contingent upon the military pressures of the Second World War for the development of radar. Intelligence testing was devised to meet military requirements in the First World War.

Suppose a message was received on the hydrogen frequency, in a code consisting of prime numbers that was translated by mathematicians. This, argues Weston (1988: 93), would be more astonishing than finding extraterrestrials: 'It would be the most startling confirmation possible of the *non*-relativity of modern scientific culture.' It would go against massive research and evidence that science, its status and meaning, is structured by its social setting.

Both Weston and Munévar see the evolution of science as akin to the evolution of a species which occupies a territorial niche. Pursuing the biological analogy Weston argues that science, like any other species, is 'conditioned by a vast number of unpredictable and for all we know unduplicated events' (ibid.). In biology it is a truism that, given similar selection conditions, many distinct life-forms will evolve in the same niche. *Homo sapiens* shares the same terrain with bacteria, mosquitoes and various other mammals. Even within *Homo sapiens* cultural variation is such that many societies will not evolve into a scientific culture. The appeal to evolutionary contingency and diversity leads both Munévar and Weston to a sceptical attitude towards SETI. It can be summarized as follows: there are no sure grounds for SETI's assumptions about ET civilizations which are capable and willing to communicate with us by radio. There is no reason to suppose that intelligent life has evolved anywhere else as it has here; that another scientific culture like ours exists anywhere; no guarantee that it would discover radio, and if it did, there is no guarantee that it would use it to communicate with us. To this, Weston adds that SETI's preoccupation with radio astronomy might even prevent us from recognizing a truly alien life-form.

But is Munévar and Weston's scepticism as damaging to SETI as it first appears? Their arguments draw attention to epistemological problems arising out of SETI's assumptions about the universality of science and radio communication. But they do not undermine the overall legitimacy of SETI or demonstrate that ETI cannot be found. A distinction between a strong and a weak sense of SETI might be helpful when assessing the damage inflicted by Munévar and Weston.

A strong interpretation of SETI involves the assumption of a universal and inevitable evolution from the primeval slime to radio astronomy in any environment similar to Earth. This model is demonstrably vulnerable to the Munévar–Weston critique.

Nevertheless, SETI might survive on a weaker interpretation of its programme. The weaker version accepts the contingency of science and radio communication. It does not have to show any commitment to beliefs about the evolution of science, and would accordingly admit much lower probability rates for any of the factors in the Drake equation. The search could continue, but with less optimism. The aliens may not be like us but if there are enough of them, perhaps just a few will be like us and develop along broadly similar lines, having encountered broadly similar problems relating to survival, notwithstanding a variety of different means of reaching a similar goal to ours. Thus a radio search may still be justified, even though it assumes a similar natural history, technology and culture to ours, which, according to Munévar, is highly unlikely. This would be a search with only a slight possibility of success, but it would have the advantage of focusing on a possible intelligence that is meaningful at our level of understanding.

SETI scientists could, however, admit that there are many contingencies in the development from prebiotic life to radio astronomy but insist that on the way there are certain threshold states which make the next stages more probable. Some stages in evolution are so important to survival that it does not matter how they are reached. The phenomenon of equifinality has been frequently observed in biology, whereby organisms reach a certain advantageous state by many different means. Once past the threshold state then other stages have a greater degree of inevitability about them. For example, once a species develops speech it has the faculty to express humour, tell stories, produce poetry and educate. Once a society has a computer the threshold towards high technology is almost inevitable.

In reply to their critics, SETI scientists could argue that the Drake equation and its analogies with the emergence of terrestrial intelligence are heuristic not probative. They could point out that, since the scientific revolution, it has been recognized that attempts to deduce the nature of physical reality from first causes are doomed, and that the case for or against SETI can only be determined experimentally.

Eventually, if not already, SETI's appeal to an experimental basis must face another line of criticism – one that is familiar to the theologians.

The Great Silence: Fermi's Paradox

Enrico Fermi, the Italian physicist, responded to the arguments in favour of ETI with his now famous retort – where are they? This has become known as Fermi's Paradox, whereby all the arguments indicate the existence of ETI but there is no evidence of their existence. Where are they, if they have had millions of years to settle here? The Earth is about 4,600 million years old, with human life emerging 2 million years ago, electricity and radio less than 200 years old, and space flight less than fifty years old. But the universe was at least 10 billion years old when the Earth was formed, which has allowed plenty of opportunity for many other civilizations to have emerged. Some of them could have been in existence for millions of years before the origins of human life. This is long enough for them to have explored the entire galaxy. They ought to have been here by now. So where are they?, asked Fermi.

Evidence of the Great Silence is the strongest case against SETI. Parallels with the theological problem of *silentium dei* are striking. Bertrand Russell was once asked what he would say if God called him to account for his atheism. Russell allegedly replied that he would respond to God by asking Him why He made evidence for His existence so poor. This seems to be the case with ETs.

Scientists with the US military have deployed a world-wide space surveillance system consisting of telescopes and sophisticated radar equipment which can track spacecraft. Every object over 10 centimetres in diameter has been catalogued, of which there are over 7,000. In addition, there are over 10 million fragments of debris, none of which can be attributed to an extraterrestrial source. So far no extraterrestrial artefacts have been discovered and no information regarding the communicative intentions of extraterrestrials has stood up to scrutiny. However, it might be argued that [1] we would not be capable of recognizing their products; and [2] we might not have found them. It only takes a few generations for a whole civilization to disappear as we know of the vanished civilizations of Asia and Africa, and cities completely obliterated by the elements within a millennium or two.

Hart's case for the uniqueness of humans

The fact that the inhabitants of planet Earth have entered the space age and contemplate colonizing voyages throughout the solar system and beyond has lent support to SETI's opponents. If we are but a few centuries away from interstellar voyages, might it not be the case that our ET neighbours are in the same position? If we are on the threshold of colonizing the galaxy – a mere million or two years away – then why haven't others, with an earlier start, done so already? This paradox was examined by M.H. Hart (1975) who argued that the obstacles to interstellar travel have been over-exaggerated and that the only reason the galaxy is not teeming with intelligent life is because we are unique. The force of Hart's argument cannot be underestimated: SETI's credibility rests on the argument that technological intelligence is widespread. If this is the case, maintains Hart, they must be here.

Hart's attempt to demolish explanations of the Great Silence is worth elaboration. First, he rejects physical explanations which claim that extraterrestrials have never arrived because 'some physical, astronomical, biological or engineering difficulty makes space travel infeasible' (Hart, 1975: 128). Space travel, he argues, does not present insurmountable problems. It may be the case that a one-way trip to Sirius at the speed of light would take eighty-eight years which is well beyond current technology. But Hart sees ways of overcoming such problems: youthful voyagers in states of suspended animation; using drugs to slow down their metabolism; future developments in biology which could make it possible to freeze and thaw out warm-blooded animals. Hopefully they would take self-functioning instruments to do this with them as there would not be anyone waiting at the end of the journey to thaw them out. Or, as Hart speculates, extraterrestrials might not be warm-blooded, thus circumventing this problem. The length of time required for the journey might not be a problem for extraterrestrials who may have life-spans lasting many thousands of years and not regard a 200-year journey as a dreary waste of one's life. Deep into the realm of speculation, Hart also considers frozen zygotes sent out in spaceships piloted by robots, or the use of techniques of time dilation. A journey lasting many generations could be mounted, as long as comfort and a well-organized social structure was part of the conception. The energy requirements pose few problems for Hart, who envisages nuclear-powered rockets scooping up oxygen as they speed through the galaxy.

Hart also rejects sociological explanations of the silence, which maintains that they choose not to come because of a lack of interest, motivation, organization, or because they suffer from a tendency to self-destruct. The weakness in these explanations, says Hart, is that they fail to consider the fact that civilizations can and do change. These explanations, he says, may only hold for one civilization at any one time. To cite them as an explanation of silence would require showing why they hold 'for every race of extraterrestrials – regardless of its biological, physiological, social or political structure – and at every stage in their history after they achieve the ability to engage in space travel' (Hart, 1975: 132). Hart dismisses as a non-testable hypothesis the argument that advanced technological societies are likely to self-destruct. There is no evidence of any doing so, he says, and the only one we can observe has, so far, not self-destructed.

SETI advocates, it would seem, cannot have it both ways. If, as SETI exponents argue, life on Earth is typical, then they are obliged to acknowledge the possibility that some, if not all, extraterrestrial civilization will have colonizing tendencies like ours. Of course they might respond that we are typical, and that just as our colonizing tendencies evaporate with the awareness of the massive costs involved and the immorality of such conquests, so it would be with extraterrestrials.

Hart's third category of explanations for the absence of extraterrestrials is classified as 'temporal explanations'. These explanations appeal to suggestions that they have not yet had enough time to reach us. To assess the plausibility of this explanation one needs to know how long it could take a civilization to reach us once it has embarked on a programme of space exploration. Hart considers expeditions from Earth to the nearest one hundred stars as a typical example. These are all twenty light years away from our Sun. Colonies might be established near them from which further voyages could be launched. Without any pause between trips Hart estimates that the galaxy would be traversed within 650,000 years (ibid.: 133). If we allow for a period between each wave of expansion of about the length of time it takes for each voyage, this would double the total number of years. But this would still mean that an advanced civilization should have reached us by now, unless they began their colonization less than 2 million years ago, says Hart.

The fourth and final explanation of absence rejected by Hart is that they have visited us but that we are not aware of it. There is no evidence that they are here now so, presumably, they were here some time ago. But if it is suggested that they came within the last 5,000 years, says Hart, a sociological theory is required to explain why no other extraterrestrials have visited since and why they do not remain here. Moreover, he adds, if it is suggested that they came and chose not to colonize, then a sociological theory is required to show why every civilization which could have colonized chose not to. Appeals to arguments which assert that they might not wish to colonize grow weaker, argues Hart (1980), as estimates for N increase. Given a tendency for SETI optimists to inflate N, then the chance of colonizing societies emerging has to be greater. In this way SETI's own arguments in favour of a galaxy teeming with ETI are rendered self-defeating.

Hart's rebuttal of explanations of absence is hard to sustain. Granted it only requires one colonizing advanced civilization to cover the galaxy within two or so million years, but there is no reason to assume that even one civilization out of billions will emulate the behaviour exhibited by some human societies over the past few thousand years of recent history. The possible fact that no benefit could outweigh the costs of large-scale colonization could apply to billions of technological civilizations. Moreover, there may be an infinite number of sociological reasons, or a combination of all four of the explanations of absence, which could account for the failure to colonize.

Despite Hart's assertion that we have no evidence of them, it is possible that the galaxy is teeming with forms of advanced extraterrestrial life, who are so advanced that we are unaware of them, yet being much more advanced, they would have no interest in revealing themselves to us.

Hart's four arguments against explanations of absence and his solipsistic conclusions rest on appeals to many unknown and as yet undiscovered scientific techniques, and assumptions that practices which have existed for a small part of human history are likely to be repeated universally.

Tipler's self-replicating colonizing probes

John von Neumann conceived of a mathematical proof that self-replicating machines could exist. All that is required is the appropriate technology. An extension of von Neumann's reasoning would be to conceive of a programme whereby robot probes could be set into motion with the purpose of colonizing the galaxy. It would take one – out of the possible millions – of civilizations to colonize the galaxy, given enough time. Self-replicating probes, like rabbits spreading across Australia, would eventually reach our corner of the galaxy. David Brin (1990: 160) cites calculations by Eric Jones of Los Alamos laboratories which suggest that an expanding sphere of settlements could fill up the galaxy within 60 million years. Frank J. Tipler, a sceptic with regard to the possibility of contact with ETI, has argued that just one self-copying probe could generate enough probes to have one near each star in the galaxy within 300 million years. Yet, argues Tipler, they have left no sign of their existence near or on the Earth. Even if there was only one other civilization, argued Tipler, albeit a technologically advanced one, they could have built self-replicating machines which could develop space travel and eventually colonize the planets. Even if their transport is less than the speed of light, argued Tipler, they could redevelop and colonize the galaxy within the age of the Earth. So why have they not contacted us already? Tipler concludes that we are alone.

Tipler's argument is familiar but, as he maintains, its force is under-appreciated. He suggests that if intelligent extraterrestrials 'did exist and possessed the technology for interstellar communication, they would already be present in our Solar System. Since they are not here, it follows that they do not exist' (Tipler, 1980: 267). Tipler maintains that 'an intelligent species with the technology for interstellar communication would necessarily develop the technology for interstellar travel, and this would automatically lead to the exploration and/or colonization of the Galaxy in less than 300 million years' (ibid.). This assumption that an intelligent, technologically communicative species will necessarily develop the means to explore and colonize the galaxy is based upon questionable grounds. In fact, Tipler appeals to a version of the principle of mediocrity (a belief that our own evolution is typical of life throughout the universe). Yet the thesis that intelligent life is universal is precisely what Tipler ultimately denies. It might be noted that whoever resorts to the principle of mediocrity, it is a weak basis for any argument. There is no evidence other than the development of some, but not all, civilizations on Earth, on which to support the conclusion that we are typical of life-forms – actual or possible – throughout the universe.

Tipler's argument is nevertheless worth exploration. He begins with the assumption, based on the predictions of certain exponents of Artificial Intelligence, that an intelligent extraterrestrial species 'will eventually develop a self-replicating universal constructor with intelligence comparable to the human level' (ibid.: 268). A universal constructor can construct anything that can be constructed, says Tipler (1995: 45) and it can make copies of itself in minute size using nano-technology. These constructors, or von Neumann machines, could then be combined with present-day rocket technology and would replicate themselves out of local resources and 'eventually colonize the Galaxy in less than 300 million years' (Tipler, 1980: 268). The cost in energy and expense would be minimal to the sending civilization who would only have to find resources for the first machine, which would be programmed to seek out construction material and replicate itself and then construct similar rockets which would transport copies to the next targets, and so on throughout the galaxy. The cost of such an interstellar probe Tipler estimates to be about US$250 billion at current prices. Improvements in design could be communicated by radio, or by means of self-learning programmes, such that future machines would be more sophisticated than their 'parents'. Tipler also envisages sophisticated machines equipped with artificial womb technology which could synthesize fertilized eggs. The probes could be programmed to construct organic beings, humans, from available raw materials. This would involve beaming human DNA information to the site at the speed of light, so that people could populate an area without having to make the journey. It would, however, have to be accepted that errors in duplication would be likely to occur, and that these would be more likely in a hostile environment, such that the 'living machines' envisaged in the original project would have evolved considerably beyond their original state during the 300 million years of galactic colonization.

While Tipler's suggestion, like the adaptive intelligence claimed on behalf of futuristic von Neumann machines, lies beyond immediate scientific horizons and is indicative of those weak philosophical arguments which appeal to incredible scientific possibilities, there is a plausible basis, at least, for Tipler's transport programme. The technology required would be little more than conventional rockets designed for one-way journeys to the nearest construction site. This could be accomplished by chemical fuel sources, although nuclear power could be developed for longer journeys. Nevertheless, such speculation is based on wildly optimistic forecasts about machine intelligence and instrument reliability.

According to Tipler, the search for extraterrestrial intelligence would have greater empirical credentials if it were directed towards the location of von Neumann probes. A radio search can miss its target: they might not be broadcasting at the time that we are listening; or we might miss them because we are tuned to the wrong frequency. But a search for probes is more determinate; either they are there or they are not. Tipler's conclusion that we are alone is based on his argument that, given an earlier start, other extraterrestrial civilizations would have set in motion a colonization process and that by now the galaxy should be teeming with von Neumann probes. Because they have not been observed, he concludes that we must be unique.

How plausible is this argument? Carl Sagan (1983) identified several flaws in Tipler's thesis and suggested limitations on galactic colonization. According to Sagan, if these machines are programmed to go on replicating, then they will not stop until the entire universe has been converted into von Neumann machines, which would then presumably cannibalise each other. It follows, argues Sagan, that these machines would be a threat to any emerging intelligent civilization and steps would have to be taken to restrict their development long before they proliferate. A prudent policy would be to prevent and disrupt their construction. If Tipler is correct, the entire universe is threatened by them and every intelligent society is in danger. If there are any other intelligent civilizations, they would take steps to limit this threat and the absence of von Neumann probes could well be due to the efforts of intelligent extraterrestrials. Measures to subdue the local population would have to accompany each probe, as it is unlikely that the inhabitants would stand idly by while the colonizers reproduced themselves. Under such circumstances, probes might be limited to unpopulated regions.

Tipler's model of colonization, which has also been criticized (Sagan 1983), is based upon the mathematical theory of island colonization developed by R.H. MacArthur and E.O. Wilson (1967). This model is not typical of all Earth populations and hence scarcely typical of populations throughout the galaxy. Sagan considers several conventional models of colonization by biological populations and he calculates that 'the establishment of galactic hegemony requires a perseverance to the task for a period of a billion years' (1983: 118). According to Sagan, it is therefore unlikely that any advanced extraterrestrial civilization would remain steadfastly committed to this task for even a fraction of that time. A civilization of merely 1,000 years into our technological future would hardly be interested in strip-mining and colonizing every planet in sight. Presumably, an advanced extraterrestrial civilization would have better things to do than imitate events in recent short-term Earth history.

However, Tipler could reply that his theory of colonization by self-replicating von Neumann machines does not require any long-term commitment. Once the first probe is launched, the process is potentially independent of the sending civilization, which may lose interest in the project or self-destruct. Moreover, it would only require the existence of one technologically capable civilization in the entire galaxy to initiate the process of colonization.

This argument becomes less plausible once it is recognized that developing and maturing civilizations are capable of recognizing past errors and of taking steps to put matters right. A civilization might come to recognize the dangerous folly of sending its probes and take steps to curtail them and even warn others about them. Moreover, there is the problem of determining the actual benefits to the sending civilization. Probes that simply reproduce themselves offer little in the way of increased wealth or security to their creators, nor do they enhance their quality of life or chance of survival.

Perhaps a less sophisticated, but nevertheless relevant response to Tipler's argument is to question the feasibility of a system of self-replicating probes. Such machines bear little relation to present technology or any predictable future developments. The credibility of von Neumann machines rests on an appeal to the analytic–synthetic dichotomy. Put simply, it involves the belief that if a project is not analytically impossible, that is, self-contradictory, then its implementation is merely a technical problem awaiting a practical, perhaps difficult, but not impossible solution. This kind of reasoning might eliminate logically incoherent proposals, but it can give respectability to a wide range of absurdities. For example, there is nothing illogical about the proposal to take a holiday on the Sun. So it is merely a technical problem regarding an adequate mode of transport and a suitable form of protection from the Sun's radiation. The absurdity of this proposal is not a matter of logic; it lies in the fact that its implementation would demand an almost infinite number of practical solutions to problems we can envisage and no doubt as many solutions to problems we have not even encountered.

Similar objections can be made against the proposals to initiate a series of self-replicating probes. As Munévar (1998: 191–2) shows, the vast physical and environmental differences among rocky planets within a stellar system such as ours, will require an almost infinite combination of factors that might affect the mining and manufacture of the basic components. The initial programme, or set of blueprints, would have to be sufficiently general and all-embracing in order to cope with an almost infinite range of environmental contingencies. Although it is not illogical, it is demonstrably absurd.

According to the arguments, these machines will approach a mineral-rich environment, alone or with other such machines, and begin to extract various materials, manufacture plastics, metals, and so on, without steel mills, processing plants or a sophisticated transport and distribution system, and then begin to construct complicated apparatus including rockets containing more probes. This is nothing short of miraculous. Using the best technology on Earth there is not even the remotest possibility of designing a machine that could go out and locate the necessary materials and manufacture a bicycle. Even if the appropriate technology were available, the problem would lie in the inability to write such general purpose programmes that would exhibit a flexible response to an incredible variety of environmental conditions. It would require an immensely complex computer, and the more complex its software, the more likely that there would be errors built into it.

Those who employ these kind of fictitious examples should be able to specify in considerable detail how such schemes can be practically implemented. Perhaps, for the sake of an experimental demonstration, a probe could be sent to the Sahara Desert or the Antarctic, where it could be given the task of replicating itself. This is extremely unlikely. Arguments which rest on appeals to better technology or hitherto unheard of technological advances, should be treated with caution. Frequently references are made to the narrow-minded prejudice of scientists who fail to appreciate future technological marvels. In this context references are made to the prominent American astronomer, Simon Newcomb, who in 1904, 'proved' that heavier-than-air flight would require too much energy to be practical. But this is misleading. Newcomb was not confronting the fantasies of science fiction; he was employing out-dated concepts against theory and observational evidence which indicated considerable knowledge of air flight.

SETI's response to Fermi's Paradox

SETI's response to the 'silence' problem has been imaginative, if not entirely convincing. There are both sceptical and pluralist explanations of the Great Silence. Sceptics argue that it means that we are unique and that our emergence is an evolutionary fluke. Sceptics also argue that the factors in the Drake equation are grossly over-estimated: there are fewer habitable zones, less planets, less life, less intelligence and less willingness to communicate than Drake and his followers believe. In short, we are either superior or unique. Pluralists have been more speculative. Maybe they came, did not see anything interesting and left without trace. Maybe they have not reached us yet, or they have been and we are the descendants. Maybe we missed intelligent life: it evolved, colonized the galaxy, and self-destructed long before humanity existed. Now we are among the very first in the second time around. Perhaps the Frank Tipler probes are really there, but are programmed not to respond until we encounter them. Maybe they came, dumped their rubbish and contaminated the Earth and out of the chemical trash emerged the seeds of prebiotic life. Thomas Gold (1960) describes this as the 'garbage hypothesis'. It is, of course, an untestable hypothesis. Maybe they are transmitting but their signals are too weak, failing to penetrate the dense gas clouds between the stars. Perhaps they have no detectable physical structures and exist as pure states of consciousness having long ago discarded their material bodies, or having rejected their planetary homes because they regard them as dirty, unhealthy places, have no wish to visit another one. Perhaps their technology evolved long ago beyond a stage where they would want anything from us. Perhaps our environment is inhospitable to the species which dominates our part of the galaxy. Perhaps, contrary to Hart and Tipler, they have no desire to colonize, no desire to migrate. Expansion is by no means inevitable. Ming Dynasty China was a leading sea power from 1405 to 1435 which then turned inwards and made it a capital offence to manufacture large sea-going junks. Advanced ETs might not suffer from any half-crazy desire to strip-mine the galaxy, running from planet to planet polluting and raiding its resources.

Strictly speaking, the basic premise in Fermi's Paradox need not be accepted. Strange as it may seem, we could argue that it has not been decisively shown that they are not here. They could very well be here – not in the sense that they are having secret meetings with the world's leaders or participating in experiments in Area 51 Research Center – but they could be sending messages which we have not yet received; they might be teleporting various 'sims' of themselves across the galaxy, but as we lack the appropriate receivers we are unaware of them. They, or their robot probes, might be mining asteroids which we have not yet looked at, or maybe they are currently colonizing the galaxy with robots or have evolved into post-organic beings but have continued further in the direction taken by our technology, and have succeeded in even greater levels of miniaturization. In which case they are too small for us to observe them.

One explanation of the silence rests on a weak interpretation of the SETI programme: it concedes that the combined fractions of f^l and f^c in the Drake equation (planets with intelligent and communicative life) are much lower than previously estimated. Drake estimated a rate of one planet in every hundred. But according to the weak interpretation it can be conceded that, even with the precursors of life, such as sugar and amino acids, the next steps towards life and then intelligence are more difficult than imagined. Perhaps some rare, as yet unknown, catalyst is required if life is to emerge. Perhaps some, as yet unknown, software is required if intelligence is to develop. It might also be conceded that intelligent life is more prone to natural disasters than previously thought. For example, the Cretaceous Tertiary event, plus evidence of four other mass extinctions found in the sedimentary record, destroyed most species of land animal over 440 kilos in weight and prohibited any potential for intelligence in the first 3 billion years of Earth life. The same thing, to even greater proportions, might occur throughout the galaxy.

Events in social history could reduce the fraction of L – the lifetime of an advanced civilization. Expanding populations, driven primarily by demographic pressure, tend to wreck their environments, as the Europeans did in their colonies. If this were repeated on an interstellar scale, it could be argued that the termination of an advanced civilization coincides with its capacity for interstellar communication and travel. The problem with this argument is that one cannot claim that civilizations will destruct before they colonize unless one can demonstrate that every single race will do so before it reaches the capacity for interstellar flight (Birch, 1990: 1). For it would only require one exception for the galaxy to be populated with ETs. There is no scientific law which prohibits colonization. It may, of course, turn out that no one wants to, but this is not supportive of a thesis which claims that no one can.

According to Hart and Tipler, given the age of the universe, there has been ample time for advanced extraterrestrials to have established colonies near or even on the Earth. The absence of these colonies is invoked to establish the claim that we are unique. Many of SETI's exponents reject this appeal to colonization. Fred Hoyle and Chandra Wickramasinghe (1978: 160) rebut the appeal to colonization with a counter-appeal to the predator–prey analogy in biology. No predator, they argue, seeks prey that cannot be obtained within its own lifetime. Consequently, colonization which takes longer than the colonizer's lifetime is biologically useless. Hence Hart and Tipler's appeal to colonization fails, because galactic colonization would require too long a commitment. Now it might be suggested that the extraterrestrials have exceptionally long life-spans, although it is improbable that they would survive for the millions of years that would be required for galactic colonization. However, the weakness of Hoyle and Wickramasinghe's objection is in the fact that human colonization projects are frequently carried out with reference to the benefits of future generations. Conquests, which take many generations, are conducted with references to historical destiny, one thousand year Reichs, with today's sacrifices balanced against the benefits to be gained by future generations.

Perhaps they have no need to colonize. For example, our Sun and nearby planets can provide enough energy and materials for the inhabitants of the Earth for several millions of years. Non-aggressive, non-expansionist civilizations consisting of pacifists might have no desire to communicate with us. Carl Sagan and Frank Drake have suggested that if a species developed life expectancy rates approaching immortality, they would be reluctant to take risks and would avoid interstellar contact out of timidity. One of Sagan's answers to Tipler is that there are two types of advanced society: aggressive and peaceful. The peaceful do not take risks, do not colonize, so we are unlikely to hear from them. But the aggressive are intrinsically unstable and are unlikely to pass through the dangerous phase that we have reached on Earth with nuclear weapons. So either they die off as the result of nuclear war and nuclear winter or they pass through the dangerous phase and become non-colonizing pacifists (Sagan, 1983). We thus arrive at the following paradox: only societies that can survive the critical nuclear weapons stage are those with the potential for colonization. But in order to survive as an advanced technological society, aggressive behaviour of this kind would have to disappear. Hence: 'the only societies long-lived enough to perform significant colonization of the Galaxy are precisely those least likely to engage in aggressive galactic imperialism' (ibid.: 120).

There is a further reason why a very advanced civilization is likely to develop pacifist tendencies. Suppose they have developed methods of combating the ageing process or have found ways of indefinitely repairing damage caused by ageing and only encounter death through accident or murder, and throughout their long lives have preserved, as Drake (Drake and Sobel, 1993: 160) suggests, 'a growing set of memories of individual experience'. At one level we can comprehend this, but the difference between them and us would be massive. They would have, says Drake, a fantastic obsession with safety, avoiding wars, accidents and even risky contacts with other worlds. It is possible that silence would be their preferred option for survival. On the other hand, they could realize that spreading their knowledge about immortality – and hence encouraging others to be equally concerned with safety – would be an alternative strategy for survival.

Intelligent extraterrestrials might very well remain silent for the same reasons as radio astronomers on Earth. Apart from an occasional brief signal, and inadvertent signals from radio and TV networks, which are very weak, the SETI radio searches are for listening, not transmitting. This is partly because intentional broadcasts are more expensive and also because of a fear that a hostile civilization might learn too much. It is possible that everyone is listening and no one is communicating. It is also possible that our failure to observe ET communication is because they have some better means of communication than electromagnetic waves.

David Brin (1990) offers a range of explanations of the Great Silence. He points out that the 250-million-year orbit around the galaxy poses major survival problems for various solar systems. When they pass the spiral arms where new stars are formed in superexplosions they are likely to be destroyed. But a very advanced civilization (for example, a Kardaschev–Dyson Type II or III, see Chapter 7) may simply leave this dangerous place taking their solar system far away. Consequently, the very advanced civilizations would be further away and

167

less likely to make contact; only the less advanced would remain and they might not have evolved the ability to communicate over long distances. Brin also notes that there may be many unforeseen disasters in the galaxy, such as huge black holes, which could destroy potentially colonizing civilizations. There may be holocausts caused by the effects of colonization such that the colonizers leave nothing behind them. But on a more friendly note, Brin suggests that it is likely that the most habitable planets – not too hot and not too cold – with ample water and oxygen, will be far less dry than ours. Hence land creatures would barely develop. In this respect we are unique. But intelligent life, such as dolphins and whales, will develop in the water, without our technology derived from the use of hand and fire, and hence an intelligence with no likelihood of reaching the stars.

A novel twist to Fermi's Paradox can be derived from a consideration of what ETIs may wish to obtain from making contact with us. This approach was outlined by T.B.H. Kuiper and M. Morris (1977) who suggested that a more highly developed civilization may place a high value on forms of knowledge or experiences that we have not even learnt to appreciate, and that this resource actually grows with time. Suppose, say Kuiper and Morris, that there is a certain threshold phase to this process before which we could produce nothing of value to an advanced civilization. If they were to make contact too soon, before we reach this threshold, and expose us to the superior knowledge held by the galactic community, 'instead of enriching the Galactic store of knowledge we would merely absorb it' (Kuiper and Morris, 1977: 620). Our capacity to produce new ideas and make a contribution to the galactic heritage would be restrained if the best human minds were 'occupied for generations digesting the technology and cultural experiences of a society advanced far beyond our own' (ibid.). Thus silence is maintained because early contact would remove the very reason for making contact.

Insofar as there are parallels between the *silentium dei* of the theologians and *silentium universe*, then a variation of the role of the Devil can be introduced. Just as the Devil is responsible for preventing communication with God, so a paranoid advanced species may be active in disrupting communication. Every few million years or so they will send down deadly probes to wipe out intelligent life. But maybe they do not act out of devilish motives. Maybe our kind of intelligence, with its destructive potential, is perceived as a deadly cancer that any really intelligent species will attempt to contain lest it spreads unchecked destroying whatever eco-system it inhabits. A criterion for superintelligence might be based on the capacity to prevent the spread of highly destructive advanced technological civilizations like ours. In that case, the fraction for L would be very small indeed.

These explanations are the stuff of science fiction. But much of science fiction exhibits an impoverished concept of intelligence, which is revealed in one of the most commonly employed explanations of the Great Silence: we have not heard from them because we have not developed the right technology. Perhaps it is not a question of having the right technology – more powerful radio telescopes and computers – but a question of not having appropriate concepts of technology and intelligence. Could we understand a concept of technology that is not geared to control of the environment and ourselves? The ability to think beyond an instrumental technology is beyond the reach of most SF authors who inevitably depict advanced ETIs with images of ever increasing control, akin to a 'Galactic Corps of Engineers' (Weston, 1988: 95). To consider a technology not geared to control is to lose sight of the notion of intelligence which is measured by its capacity for control. A billion-year-old civilization, says Weston, might not even value control, but 'view nature in such a way that precludes engineering' (ibid.). But even closer to home, an instrumental concept of intelligence fails to capture human intelligence, as the artificial intelligentsia have learned. A computer can generate a military strategy that could do the work of ten generals but not know how and when to write an apology.

There is a danger in too much speculation about ET life and ET intentions: in the absence of factual evidence, speculation may not only be widespread, it may take on a form of wish-fulfilment representing deeply felt psychological and even moral beliefs. This may partially explain speculations about benevolent supercivilizations who are keeping a paternalistic eye on us. But there are rules to govern speculation: it must be contained within the boundaries of scientific theory. This allows one to pursue an idea beyond the limits of so-called hard-headed, all facts-and-no-nonsense-thinking, and indeed conform with the philosophical imperative to follow the argument wherever it leads, but it is necessary to remain within the plausible extensions of currently defensible theory.

One answer to the question of the Great Silence is akin to the theologians' answer to the question of *silentium dei*: it requires a transformation of our concepts of intelligence and our purpose in communicating. Maybe human intelligence is not the ultimate product of evolution.

Are they among the asteroids?

Many explanations of the silence assume that the extraterrestrials simply choose not to colonize. But B. Zuckerman (1985) suggests that a very advanced and long-established technological civilization of several million years might have little choice but stellar migration. Zuckerman's argument adds a new dimension to Fermi's Paradox in which doubt is cast upon the SETI community's preference for radio searches over search for evidence of migration. According to Zuckerman, several million-year-old supercivilizations would be obliged to consider mass migration as their sun becomes a red giant. If this is the case we should expect the galaxy to be teeming with extraterrestrials, and evidence of their existence would be derived from their expeditionary ventures rather than radio signals.

Zuckerman's argument, like Hart and Tipler's, suggests that advanced ETs should have colonized the galaxy by now. The fact that they have not is taken as proof that we are unique. Michael D. Papagiannis expressed the now familiar dilemma as follows:

> Either the entire Galaxy is teeming with intelligent life and hence our Solar System must have been colonized hundreds of millions of years ago, or else there are no other inhabitants in our Solar System and hence most probably neither anywhere else in the Galaxy.
>
> (1978: 277)

Before accepting the 'bleak' conclusions that we must be alone, Papagiannis suggests a search in the asteroid belt, which could be an advantageous place for a galactic society living in space colonies. Papagiannis reaches the suggestion for a search in the asteroid belt from his considerations of six arguments employed in support of the case for our uniqueness: first, life expands to occupy all possible living space; second, interstellar travel is theoretically and technically possible, and colonies could be built as stepping stones thus making star trips possible for future generations who have lost all emotional ties with Earth. Third, even without major new discoveries, with steady technological progress, stellar missions could be undertaken within a few centuries (ibid.: 278). Fourth, once the interstellar threshold is crossed, it would only take a few million years for the whole galaxy to be colonized. Fifth, the attractive features of our solar system, with its hot stable star, would not be missed by colonizers. Sixth, the colonizers would be accustomed to space living and might not need to live on a habitable planet like Earth, but prefer to live in space colonies.

It can be deduced from the first five arguments that either they are already in our solar system or we are alone (ibid.). Their apparent absence suggests the latter. But the sixth argument provides Papagiannis with an alternative. As they are very likely to prefer to live in space colonies, they might have selected the asteroid belt as a location for their colony. They would have access to natural resources by mining the mineral rich asteroids. They would be close enough to the Sun in order to harness its energy. It may even be the case that many fragmental asteroids are the result of their mining projects, adds Papagiannis, who suggests a thorough search of the asteroid belt before accepting the conclusion that we are unique.

Of course identification of extraterrestrial activity would be difficult; from this distance natural asteroids and colonies would be indistinguishable. So Papagiannis suggests that we look for radio leakage, infrared observation for unusual temperatures, and space missions designed to seek reliable photographic evidence.

There is still the question regarding their apparent silence. Why do they remain silent when they are so near and contact would be relatively simple? Papagiannis suggests that they suffer from 'confusion and indecision'. For centuries they have observed us and seen no sign of technology. Yet during the past fifty years it has grown at an unprecedented rate. Their problem is whether to crush us or help us but, according to Papagiannis, for the moment they have postponed any decision as they are aware that, for the present, we cannot find them.

The Zoo Hypothesis

Another explanation of the silence is the 'Zoo Hypothesis', proposed by J.A. Ball (1980: 242), who said that, 'The perfect zoo ... would be one in which the fauna inside do not interact with, and are unaware of, their zookeepers.' This hypothesis suggests that benevolent super-beings look on us as a species in quarantine, at present too dirty and dangerous for contact, or that the Inter-galactic Council has designated Earth as a nature reserve. Perhaps, also, if they are very advanced they do not consider us worth contacting. The Zoo Hypothesis has a number of versions, most of which are falsifiable but not confirmable. As Ball put it: 'We shall never find them because they do not want to be found and they have the technological ability to ensure this. Thus this hypothesis is falsifiable, but not, in principle, confirmable by future observations' (ibid.: 243). Intelligent ETs may have penetrated the solar system with their probes and be keeping a low profile because of an embargo on contact with any inhabitants who are still planet-bound. This is to avoid lending any encouragement to plans to quit their planets prematurely, as first they must somehow demonstrate their fitness to mix with other beings. In such a case, silence would not mean absence, but an unwillingness to speak. According to Papagiannis (1988), even if no message is heard after the next million hours of search, we should not conclude that we are alone. Communicating civilizations may have an ethical rule whereby newcomers are obliged to pass an entrance test; for example, avoidance of overpopulation, disease, global war or environmental disaster. He concludes that a negative result of a long and comprehensive search need not simply mean that we are alone in the galaxy, but that we are not ready to join a sophisticated 'Galactic Club' which has rules for membership, and the headquarters of our galactic region has not yet issued an instruction for us to be admitted. But if we can solve our problems then we may, in time, receive an invitation. Papagiannis concludes that we need not waste efforts on expensive searches but should just concentrate on our own problems and meanwhile try to eavesdrop on them while we wait.

Critics of the Zoo Hypothesis draw attention to its inherent untestability and its failure to explain why the Earth and its neighbourhood were quarantined before life appeared on it. But most critics point out that it requires a total and an almost incredible degree of unanimity throughout the entire galaxy. Given the problem of travel and information transfer – even at the speed of light – it would seem that a well co-ordinated policy regarding Earth's quarantine would be practically impossible to implement and maintain. It is, say critics, impossible to maintain social cohesion over vast distances when the message, sent out at the speed of light, would take 100,000 years to cross the galaxy. Thus an edict, 'Leave Earth alone!', would take 100,000 years; a reply, 'Why?', would take another 100,000 years; and 'Because ... ' would be another 100,000.

An answer to this problem of social cohesion through communication has been proposed by I.A. Crawford (1995) who argues that faster-than-light (FTL) travel is not impossible. Against the widespread view that Einstein's general theory of relativity prohibits FTL travel, Crawford cites numerous sources which suggest that it is not incompatible with current interpretations of general relativity. Crawford's point is that no theory actually prohibits FTL travel; we simply do not know whether it is possible or not. Crawford, however, acknowledges the paradoxes involved with time travel – for example, travelling back in time and assassinating one's grandparents – but he insists that providing that the causality violation (associated with time travel) does not occur, there are no theoretical objections. Faster-than-light travel, he argues, need not involve time travel and there may well be limits to time travel which permits FTL travel, such as tachyon drive (the conversion of ordinary matter into tachyons and back again), wormholes (short cuts through space and time), and a form of warp drive involving a distortion of space-time ahead and behind a space vehicle.

One must be wary of arguments which appeal to theoretical solutions which have no more substance than the fact that they are compatible with extensions of prevailing theories. However, Crawford does acknowledge that no one has any idea how to achieve FTL travel in practice: 'No one has any idea how to build a warp drive, or construct a wormhole or turn a space-ship and its crew into tachyons' (1995: 211). But it is theoretically possible, that is, permitted by the laws of physics as currently understood. The problems occur when the issue is classified as a technical problem which, of course, can be solved by any sufficiently advanced civilization. For these societies can do anything that does not contradict established theory. This is a convenient way around any problem.

Merely because relativity theory does not forbid something does not make it possible or remotely likely. Without some inkling as to how FTL travel is to be achieved, appeals to what is possible 'in principle' carry little weight. Permission does not make things possible. But as a piece of reasoning in the context of discovery, such an argument indicates where further investigation could be undertaken by opening up the scope of permissible research.

But let us stay with this argument. If FTL travel and communication are possible, then galactic colonization is also a possibility. This would seem to support Tipler and Hart's view of our uniqueness: if ETs can travel or disperse information faster than light and are not here now, then it would suggest that we are unique. However, FTL travel and communication offer another possibility. They would facilitate more social coherence over a larger area, giving galaxy-wide political controls which could maintain a policy of non-interference, thus removing one of the main objections to the Zoo Hypothesis. Thus Crawford concludes, that 'the Zoo Hypothesis can only be a valid explanation of the apparent absence of extraterrestrials if FTL travel is allowed' (ibid.: 216). Hence the apparent absence of ETI means either that we are unique, or that ETs, having discovered how to travel or communicate faster than light, can maintain a policy of non-interference.

One version of the Zoo Hypothesis merits further consideration, as it is a testable, if slightly implausible, response to Fermi's Paradox. It is known as the 'Peer Hypothesis' and was developed by Paul Birch (1990).

The Peer Hypothesis

Birch hypothesizes a race of superETIs with no limits on their technology, who can travel from galaxy to galaxy, and can build with planets, black holes and stars, and manipulate the topology of space-time. These superbeings can manufacture their own space-times, create whole universes, and control them as we would control a small zoo. Could it be, asks Birch, that we actually live in such an engineered universe? The intelligence gap between these superETIs and us would be immense, no less than the gap theologians pose between God, who comprehends all, and mere mortals. Assuming only one creative intelligence behind the universe, it is likely that all intelligence made in its image will be roughly similar, argues Birch. Thus according to the Peer Hypothesis, 'Many ETIs have been created, but their development has been deliberately arranged and guided so as to be roughly contemporaneous with our own' (Birch, 1990: 3). This hypothesis allows for the SF scenario of thousands of ETI civilizations throughout the universe, but it predicts that Earth is neither advanced nor backward, but probably typical of the development of intelligence throughout the universe. Birch suggests that these broad similarities would be because the creators want a universe that would be eventful and interesting, as 'history is at its most diverting when many groups interact, and when their capacities and interests are neither wildly disparate nor absolutely identical' (ibid.). Of course, there may be many different reasons behind the creation of such a universe: intellectual challenge, entertainment, and so on.

If the range of development of intelligence and technology is roughly similar throughout the entire universe, there would be enough diversity to enable different cultures to learn from each other after first contact, but not too much so as to preclude each other's cognitive thresholds. Fermi's Paradox would be resolved as, being similar to us, ETIs would not yet possess the technology required for colonization. But like us they will possess it soon, and then they will come. Moreover, after first contact others will follow soon, as the technological skills will not be that far apart. According to Birch, there are military implications with the Peer Hypothesis. Our peers would not be so powerful so as to easily defeat us, or too weak to be insignificant. Being similar to us they would come prepared and war should be considered a probable outcome.

How should we assess the Peer Hypothesis? Is it mainly a piece of imaginative SF and at variance with conventional theories regarding the origin of the universe? It requires a race possessing superintelligence with the power to create multiple universes. However far-fetched this may seem, it is not incompatible with theories concerning the origin of the universe and other universes too. John Gribbin (1994) has advanced the view that entire universes can be created out of black holes where matter has been squeezed to the point of becoming a singularity. When the singularity begins to expand, a new universe can be created which will exist in its own space-time dimensions. He suggests that there could be numerous such universes. Gribbin also considers that the squeezing operation could be an act of intention.

> Bizarre though it may seem, if you could squeeze a kilogram of butter (or anything else) hard enough to make a black hole, that black hole could be the seed of a new universe as big as, or bigger than, our own.
>
> (Gribbin, 1994: 250)

If this were possible, a superintelligence may well have created several universes along the lines suggested by the Peer Hypothesis. As a matter of fact, multi-universe production may well be within our technological capability. According to Gribbin:

> The technology is not so far-fetched, and would involve a super-powerful hydrogen bomb explosion somewhere in space at a safe distance from the Earth. It is even conceivable that our Universe was manufactured deliberately in this way as part of a scientific experiment by a technologically advanced race in another universe.
>
> (ibid.)

Of course, the postulation of a race of superextraterrestrials, who can create and manipulate space-time, only raises further questions: where did they arrive from? Is there a race of supergeniuses behind them? Nevertheless, the Peer Hypothesis answers Fermi's Paradox, and is consistent with evidence pertaining to the existence or non-existence of intelligence throughout the universe. But is it true? Is it likely? At least the hypothesis allows for these questions to be determined by evidential considerations and predictions. Given mankind's present ability to communicate to the stars and the likelihood of interstellar travel by 2200 AD, then assuming Earth is typical of other habitable regions in the universe, with a similarly developing timescale, we should expect to meet our peers fairly soon.

Why come here when you can travel by information transfer?

Most versions of Fermi's Paradox appeal to the likelihood of galactic colonization. Given the age of the galaxy, there has been ample opportunity for at least one advanced civilization to have colonized the galaxy. Hart (1975), as we have seen, rests his case for uniqueness on the fact that not even one colonizing civilization has reached the Earth despite ample time to do so. Objectors have pointed to the drawbacks of colonization which include the cost of highly expensive physical interstellar travel. But as the projected number of advanced civilizations is high and as Zuckerman (1985) argues, mass migration might be the only option for long-established civilizations facing the collapse of their sun, the evidence of colonization should be available. So where are they? This problem has been given a new solution by Louis K. Scheffer (1994), who sees a resolution of Fermi's Paradox lying in an appeal to developments in machine intelligence.

Scheffer considers alternative methods of travelling across the galaxy other than expensive physical craft, and offers some intriguing proposals for downloading our consciousness into intergalactic cyberspace. He begins with the notion of teleportation which, as yet, has been confined to science fiction. The idea is that a sufficient level of data is collected and measured from particular persons, including their mental states, and then these data are transmitted to their destination via electromagnetic signals. There have been many objections to this proposal. One objection was that the very act of measuring the original object would collapse its quantum state and prevent an exact replica from being constructed at the destination. Scheffer drew attention to a theoretical counter to this objection by C.H. Bennett *et al.* (1993) who suggested that by pre-arranging the transfer of correlating but unmeasured particles the exact quantum state could be reproduced at the destination despite collapse of the original quantum state. But this theoretical solution, says Scheffer (1994: 158), does not resolve a major problem of teleportation, namely that the reproduction of any object of significant size would require sending an enormous amount of information at great expense.

Thus Scheffer suggests a less expensive method involving information transfer by computer program. The questions he then poses are [1] how does one get the receiving computer to its destination in the first place?; and [2] is it possible to construct a conscious being as a computer program? The answer to the first question is that the teleporting civilization sends out probes to various destinations which will contain robots that will build and maintain receiving computers. As far as the second problem is concerned, Scheffer considers the notion that a consciousness could be mapped and teleported by computer. Most scientists, he says, would in theory agree that the state of every sub-atomic particle of any object, including a human being, could be measured and its evolution under quantum mechanical rules could be simulated. Then, in theory, its behaviour could be predicted. This, of course, is a massive problem. A human body contains at least 10^{27} atoms and even one bit of information per atom is too much information for any foreseeable computer to handle. Moreover, as Scheffer concedes (ibid.: 159), we have no idea how to measure a human being on a sub-atomic scale.

This problem has been addressed by Lawrence Krauss (1996) who points out that if it took just one page to describe the configuration of each one of a person's atoms, its electrons, its orbits, state of motion, its nearest neighbours, etc., then to get all the information for a whole person on 10-gigabyte hard drives would require a stack reaching from Earth to one-third of the centre of the Milky Way. Using the best of current technology it would take longer than the age of the universe to download the information.

Yet despite the impossibility of practical implementation, Scheffer considers teleportation as a thought experiment. The problem might be reduced if we scale down to the transfer of a simulated nervous system, a brain whose operations might be simulated by a computer program. This again raises major philosophical problems, which Scheffer recognizes, as it can be argued that the program would not be a real consciousness, but only a simulation (see Searle, 1987). Nevertheless, appealing to the most optimistic predictions in artificial intelligence research, Scheffer concedes that the reason why a complete simulation is not possible today is largely because we know so little about the human brain. Thus, for the purpose of sustaining the thought experiment, Scheffer assumes that 'in some cases shorter conscious programs are possible'. Despite reservations with this thesis, namely, the objection that no increase in the perfection of a simulation will transform its identity to the original object, we shall continue with Scheffer's thought experiment which could suggest a plausible solution to Fermi's Paradox.

Scheffer proposes that once consciousness is captured on a computer program then its transfer would be as follows: a computer program is started on one machine, then it is stopped and the current state of the program is sent to another machine at a different location, which then resumes the execution of the program. It would appear that the program – a conscious entity – has been teleported. To secure plausibility Scheffer cites work in today's computer networks which already allows computers to send incomplete programs which are devised in such a way to allow another machine to complete it. There are also developments in cyberspace technology where 'simulations' of conscious experience could be teleported. The fundamental philosophical objection is that if a simulation of a person is teleported the person remains behind, unless we revise our views which place a conceptual restraint on the idea of a person being in two places at the same time.

So how does Scheffer's thought experiment resolve Fermi's Paradox? Teleportation eliminates the need for physical systems of transport and colonization, so they are not likely to be here in any physical sense even though their computer simulations are hurtling around the galaxy. Once the first civilization developed the technical means of information travel its members can go anywhere at the speed of light at a fraction of the cost of physical transport. It requires no major engineering challenge, no need for large spaceships or programs for strip-mining the galaxy. Now one of the arguments in support of Fermi's Paradox is the large numbers argument which Hart (1975) employs when he says that only one civilization out of billions will need to be colonizers. Opponents are then required to show why none, out of billions of civilizations, chose to colonize. Scheffer's answer is that once one civilization has adopted a system of information transfer then others will have a disincentive to engage in physical travel, as it will be cheaper to join the existing galactic teleportation travel club. Moreover, Scheffer speculates, if the club offers easy galaxy-wide teleportation, there is a likelihood that it could lead to a single culture throughout the entire galaxy, and if this culture has decided against colonization of the Earth, then the large numbers argument loses its force.

Similar arguments can be applied with reference to very advanced civilizations facing extinction; migration could be impractical and risky, but if they have developed sophisticated techniques for electronic communication their consciousnesses could be teleported across the galaxy without any obvious signs of mechanical transport. They could be here and we have not yet made contact.

Scheffer's argument can also be adapted as a rebuttal of one of the arguments employed against the Zoo Hypothesis. For example, it is frequently argued that the Zoo Hypothesis requires an enforced agreement among millions of civilizations that Earth will not be colonized or visited. This, it is claimed, is impossible to enforce. But an homogenous galactic culture could enforce a code of practice in which Earth is treated as a nature reserve.

What has Scheffer established with his theory of teleportation by information transfer? In a series of steps he concluded that [1] travel by information transfer is plausible; [2] that it is cheaper than physical travel over interstellar distances; and [3] that if there is intelligent life in the galaxy using interstellar information travel, it has very likely facilitated the emergence of one large civilization. The first step, however, encounters philosophical objections concerning the possible simulation of consciousness; the second step is clearly acceptable if practical and conceptual objections to the first can be overcome; while the third step is highly contingent. More communication, contact and travel of whatever form, need not lead to homogeneity: it could equally produce exaggerated diversity. Despite networks of almost instant communication the warring nations of Earth are far from homogenous.

What observational predictions can be derived from Scheffer's thesis? We could eavesdrop on their communication network if they are transmitting at a range within our reach. But, says Scheffer, they would probably transmit at between 50–60 GHz range which would be cheaper for them, but difficult for us to detect from the ground. They may have sent ships to various places in order to repair or replace damaged computers. But this might only require one visit per stellar system, which means that we would be unlikely to find them. In any case, they could have perfected their network before the Earth existed and have no further need for physical travel.

These considerations allow Scheffer to reverse Fermi's Paradox: if humans are typical of galactic life and such intelligent life has evolved many times in the galaxy, and since we have not been colonized then there must be some reason why. Scheffer's answer is that there is already a large unified galactic society, where there is no incentive for physical travel, where there is no need to trade or transport material goods.

A note on information transfer

The idea of a society, whether it is terrestrial or intergalactical, within which all relationships are conducted by means of information transfer, raises profound philosophical problems. Electronic messages, broadcasting, the Internet, in their various ways, provide a substitute for physical interaction. But these systems evolved against a background of physical contact. Politicians, scientists and members of the business community, may communicate by means of electronic data exchange, which substitutes in part for a physical presence. Major issues are still addressed in meetings and conferences where participants meet each other in the flesh. This is not a lingering habit from the days before the communications revolution: physical presence is fundamental to meaningful interaction between intelligent beings and substitution leads to a diminution of meaning.

Information transfer is precisely what the expression says: an exchange of information which already possesses embodied meanings bound up with the physical interactions of persons. Whether the distances are long or short, information transfer cannot amount to a conversation, which requires perpetual learning, adjustment and a degree of flexibility which are not communicated via a printer or telescreen.

Moreover, electronic communication is not merely parasitic upon embodied communication; it is also parasitic upon the material production of goods and wealth which is a direct outcome of embodied communication and a direct relationship with physical reality. The electronic community is a bogus community which exists parasitically on real wealth-creating communities. Electronic communicators can transfer wealth across the globe within a fraction of a second, but it is wealth that others create. They are consumers, not producers, of wealth. The so-called information revolution is a revolution in consumption, offering nothing towards the production of wealth. Real communities are bankrupting themselves in order to support an elite communicative system and its expensive hardware. Membership of an intergalactic network – if one exists – would hardly be advantageous if it is like the terrestrial Internet, consuming time and resources (costly electronic hardware, miles of fibre optic cables) and producing an exchange of addictive stream of consciousness babble, not conversation. Critics of the cyber-culture have argued that many terrestrial electronic networks provide a shallow form of community; pseudo communication, a culture consisting of aggregates of electronically linked individuals, a disembodied society (see Sardar and Ravetz, 1996).

Proponents of a cyberspace culture have claimed that their disembodied system of communication is akin to an exchange of pure Platonic forms. Cyberspace technology, so it is claimed, offers a release from the enslavement of our physical bodies to the world of forms. Michael Heine (1991: 64), for example, sees cyberspace in terms of the Platonic world of forms, the world of 'inFORMation':

> Cyberspace is Platonism as a working product. The cybernaut seated before us, strapped into sensory input devices, appears to be, and is indeed, lost to this world. Suspended in computer space, the cybernaut leaves the prison of the body and emerges in a world of digital sensation.

This imagery invokes Plato's cave allegory, according to which the prisoners, when released from their fictitious existence in the cave, ascend to the realm of active thought. In Heine's example, however, the prisoners simply don the sensory apparatus attached to the computer and engage in the world of inFORMation. Such arguments omit something of greater significance in Plato's thought. For Plato described an artificial world where prisoners chained together in a cave, unable to interact with the physical world and each other, would project their fantasies onto the shadows on the walls of their prison. Unable to move, interact with each other and the world around them, they were unable to test whether their fantasies corresponded with reality and were consequently doomed to a life of ignorance. The cybernaut, strapped to sensory input devices is in many respects more akin to the enslaved prisoners in the cave, unable to interact with anything beyond the information provided by the computer. Participation in a world of active thought requires interaction with embodied minds, physical objects, in the real world. Like the inhabitants of Plato's cave, electronic communicators will require to move in a real world lest their fantasies overtake them. Information transfer without travel may lead to a system of knowledge – undoubtedly linked to a system of political authority – wherein the truths of the galaxy are as worthless as the truths of the cave. If contact is made with ETI, communication imitative of our terrestrial modem culture is to be avoided at all costs.

Is Fermi's Paradox due to the wrong model of galactic colonization?

According to David. G. Stephenson (1980), Fermi's Paradox is a product of a particular model of galactic colonization. A different model, he argues, need not produce a paradox whereby the galaxy is teeming with intelligent life of which there is no evidence within the solar system. Most models of galactic colonization, says Stephenson (1980: 248), assume that interstellar colonies will have a life cycle paralleling that of a virus, where the 'virus or an interstellar vessel, travel through an alien medium with most of the information in a dormant state and being irrelevant to the surrounding conditions'. When it reaches a suitable place the information is activated and put to use in its new environment. The human equivalent of this model is the eighteenth- and nineteenth-century European emigration to the USA where the immigrants took with them skills and information which were of no value on the journey, but were essential in modifying the New World in order to replicate some of the conditions they had left behind in Europe. The 'virus' model presents an image of colonists travelling from planet to planet in a state of limbo but then on arrival setting up colonies similar to the home environment, or at least seeking out planets offering a similar environment to the one they have left. This model is reinforced with accounts of the vast distances in interstellar space, which suggest that the travellers live very long lives or have acquired techniques of hibernation. Once this model is adopted, the absence of colonies can then be cited in support of claims regarding human uniqueness. But Stephenson considers an alternative model which avoids Fermi's Paradox.

Instead of a passive journey where the vessel is merely a means of transporting beings with a dormant capacity to activate information on arrival, Stephenson's model depicts the vessel as an equivalent to a higher organism in a constant state of activity, with its information relevant and in use. In short, the vessel would be a travelling biosphere, a permanent ark, in which its inhabitants are completely integrated. The vessel might even grow on the journey, drawing material on the way. It would not be a mere means of transporting beings from embarkation to destination, it would be the home of its inhabitants, adapting to conditions wherever it went. Not requiring settlements at either end of the journey, the human analogy is a nomadic tribe that has adapted to an environment that fulfils its needs as long as they are mobile. Such a community would not require planetary colonies and would be unlikely to visit us. There could be, suggests Stephenson, thousands of interstellar vehicles of this type throughout space and undetected.

Now Fermi's Paradox arose because the 'virus' model of colonization suggested that extraterrestrials, like eighteenth- and nineteenth-century Europeans, would inevitably end up in the New World after populating all corners of the Earth-galaxy. But a galaxy teeming with interstellar vehicles which are independent homes would not lead to the colonization of the Earth.

Are there any observable predictions in Stephenson's thesis? Stephenson's proposal not only avoids Fermi's Paradox, it also predicts likely zones for the detection of their vessels. Thus in our solar system, he suggests (1980: 248–9),

[T]he search should concentrate on the outermost bodies of the Solar System as being the closest to the interstellar space. It is an interesting speculation that an unusual orbit and planetary parameters of Pluto might, in part, be due to the activities of visiting interstellar craft that have reduced the body from a conventional outer planet to a cosmic slag tip of unwanted elements.

(ibid.: 249)

Conclusion

So much for the Great Silence. Perhaps it is too early to draw a curtain over this debate. It is only this century that speculators about the existence of ETI have been given the technology to conduct an empirical search for it. At the present time so little of the universe has been explored, and so little (less than 1 per cent) of the solar system. There is no firm knowledge of how life began on Earth, no *a priori* knowledge to rule out claims to how it might have begun elsewhere. It is too premature to speak of a paradox in the context of the Great Silence, as yet there is not enough evidence either way. 'Absence of evidence is not evidence of absence.' As long as the number of negative searches cannot be regarded as a statistically representative sample, all conclusions regarding the absence of ETI remain unsupported. Either side could be right. The issue is undetermined. One cannot base predictions about the existence or non-existence of ETI on hypotheses regarding their alleged activities. There is no alternative other than a comprehensive search. When speculation reigns, the priority of observational research must reassert itself in the scientific domain. Fermi's Paradox is employed as an argument that research into the possibility of ETI should not be funded or undertaken. It has no force against a commitment to observational science. But this is not to rule out the possibility that, if negative searches were to continue, it would begin to look as if we have missed the chance of finding intelligent life in the galaxy.

7

CONTACT WITH ADVANCED SUPERCIVILIZATIONS

> Someday, from somewhere out among the stars, will come the answers to many of the oldest, most important and most exciting questions mankind has asked.
>
> (Frank Drake, Preface to Project Cyclops, Oliver and Billingham, 1973)

Introduction

The dream shared by SETI researchers is that one day contact will be made with an advanced civilization and that such contact will provide great cultural and scientific rewards. But what does it mean to speak of an advanced civilization? Would we recognize one if we saw it? And would contact be necessarily beneficial?

Kardaschev–Dyson supercivilizations

During the 1960s there was considerable speculation concerning the evidence of supercivilizations with an immense technological superiority to ours. Nikolai S. Kardaschev, the astronomer who mounted the first Soviet ET search in 1963, published in 1964 his prediction of supercivilizations (SCs) which are billions of years ahead of us in their technology. According to Kardaschev, they would be capable of harnessing all the energy from their sun, while we at a very primitive level of evolution only take what happens to shine on our planet. But taking the argument a step further he argued that a very advanced SC could harness the entire energy output of its galaxy (Kardaschev, 1964). He suggested that the measure of a civilization's technological level would be in proportion to its use of energy, and proposed that three types of advanced civilizations could be considered. Type I would have a similar technological level to Earth, using 6.6×10^{12} watts. This civilization could engage in something akin to the present power output of Earth for the purpose of interstellar communication. Type I civilizations would have the power to restructure entire planets.

A Type II civilization would be capable of tapping its entire stellar energy, using about 4×10^{26} watts. It could restructure an entire solar system; but such a level of technology, argued Kardaschev, would have reached beyond our technological horizons. Their civilization would also have passed through the critical self-destruct phase. If they thought that communication with Earth was desirable, they could communicate from their nearest galactic neighbours.

A Type III civilization would be capable of controlling and reconstructing whole galaxies, with an energy output of 4×10^{37} watts, and could communicate across the entire universe. If Type II and III civilizations exist, their energy consumption would make them detectable and proposals were made by Sagan and Kardaschev for a search for Types II and III among the nearest galaxies.

Whereas communication with a Type I civilization might be within our communicative horizon, it may be too far away for us to receive their signal. Kardaschev therefore assumed that we are less likely to detect a Type I civilization than a Type II or a Type III which means that contact is likely to be with those who are considerably more advanced than us. Types II and III could reach us with their signal but Type III are well beyond our communicative horizon and it is unlikely that they would be interested in talking to us, although a few might wish to engage in some form of antiquarian dialogue.

A similar hypothesis was developed during the 1960s by Freeman J. Dyson (1960) who claimed that an advanced civilization might be driven by pressures of population growth forcing it to seek more living space and thus reassemble planetary matter to construct a huge sphere around its home star in order to intercept and utilize radiation from the parent star or build a habitable swarm of smaller spheres around the parent star. According to Dyson, Malthusian pressures would drive a civilization to an efficient exploitation of all their resources within a few thousand years of entering the stage of industrial development. Thus they might re-arrange the total mass of a planet the size of Jupiter to construct a spherical shell around the Sun at twice the Earth's distance from it. It would be about 2–3 metres thick and would contain all the energy from the Sun and provide a habitable biosphere for an expanding population and all the machinery for exploiting solar radiation on the inside. Such feats of engineering, he claimed, would leave indications of waste heat that we could observe in the form of infrared emissions across the galaxy.

These spheres, known as 'Dyson spheres', have actually been sought by radio astronomers. Dyson (1960) suggested searches for dark objects with a size similar to the Earth's orbit, having a surface temperature of 200°K to 300°K, and radiation in the infrared, around 10 microns wavelength. This search for 'infrared' stars could be conducted independently or in connection with radio searches.

Dyson's hypothesis, although strikingly bold, has the merit of being related to possible observations and is suggestive of possible places to look for intelligent ET life. It also has the advantage of not employing the assumption that advanced extraterrestrials are actually trying to communicate with us or colonize us. His suggestion involves an alternative to radio searches: that attention should focus upon the amount of infrared emission coming from the extended habitable zone, thus eliminating the need for an expensive array of antennae as indicated in Project Cyclops; and he therefore argues that a rational approach to SETI would be based on an extension of infrared astronomy.

Dyson actually suggests that a failure to detect these spheres should be considered as evidence of the non-existence of ETI. This may be in conformity with the falsifiability criterion of Popperian methodology but it is restrictive and could prematurely prohibit many fruitful hypotheses concerning ET life. First, ETs may not develop in directions which would lend themselves to the construction of spheres; second, there are insufficient probabilistic reasons for inferring the existence of Dyson spheres. If they do exist, however, a civilization that could construct a Dyson sphere would find the construction of a powerful communicative beacon mere child's play. Yet no beacon has been detected. While it is intriguing to admit the plausibility of Dyson spheres, the case for their existence is not probable enough to base an entire search strategy on them. There are no fine details concerning the assembly of the orbiting structures or how they are to be made habitable. In this respect Dyson's account of possible advanced civilizations suffers from the inadequacies of so many appeals to the future and faraway. We are not told in any great detail as to how their technology could solve problems which are beyond the range of our knowledge. Appeals to the current infancy of science can never be trusted. There are limits to what is technologically possible and it is the scientists' job to point them out and to show how, if it is possible, they can be surpassed.

Dyson would also have to resolve the problem of whether civilizations would be destroyed through the effects of over-population before they developed the technology for successful stellar manipulation or migration. Then there are problems concerning the observation of Dyson spheres. The main problem here is that of separating data indicative of Dyson spheres from other naturally occurring infrared astronomical objects. Detection would not be easy, as the excess infrared radiation from the spheres would be indistinguishable from that produced by a lot of dust around a star. Many infrared sources were picked up by NASA's Infrared Astronomical Satellite (IRAS) in 1983, but they could have been of natural origin, such as stars with disks of dust. The IRAS data include 50,000 sources of infrared radiation which could be Dyson spheres, although they could be baryonic matter, and as Learned et al. (1994: 327–8) point out, 'There is no reason to believe that we would have as yet identified a Dyson shell even if very large numbers are nearby in the Galaxy.'

IRAS surveyed 130,375 infrared sources that corresponded to stars. From this Professor Jun Jugaku of Tokai University selected 594 stars which resemble our Sun and searched for additional infrared radiation. Only three had an infrared excess and after further examination this was explained by natural causes (Heidmann, 1995: xviii–xix). While many objects have turned out to have the expected spectrum of Dyson spheres, there has been no positive proof, as they may well be newly formed stars. The usual method in scientific investigation is to exhaust all natural explanations before considering the unnatural ones and, even if we found a star with an excess of radiation and no immediate explanation in terms of natural phenomena, it is highly unlikely that we would conclude – without other evidence – that it indicated intelligence.

It can be argued that very advanced SCs would not be so prolific with their energy sources and would consequently remain undetected. There are also dubious assumptions about the aspirations of SCs in the Kardaschev–Dyson theory. The SC hypothesis is based on the assumption that the main aspiration of an intelligent civilization is the collection of information about the surrounding universe. This requires increasing amounts of energy. This is why there are SCs of Type I, who have mastered their planetary energy, Type II, the energy of their star, and Type III, the energy of their galaxy. But this hypothesis assumes a concept of progress which might be unique to human culture. Moreover, the concept of linear progress has not always been uppermost among human civilizations. The notion of progress took root in Western religion and was secularized in nineteenth-century biology, despite its rejection by contemporary neo-Darwinism. Contemporary science is, however, structurally dependent upon the notion of progress, but this may be accidental.

What are the benefits of contact with an advanced civilization?

Very few philosophers have addressed the question of contact with an advanced civilization, although it is widely assumed that they would be more scientifically advanced than us, and that scientists in particular and humanity in general could benefit from their superior knowledge. Thomas R. McDonough (1987: 1) argues that intelligent ETs are likely to have older civilizations, and he offers an optimistic forecast of contact: 'They have most likely faced the same problems of war, disease, starvation and pollution that plague humanity today. Any civilization that survived very long has most likely solved these problems.' If the Kardaschev–Dyson hypothesis of SCs, billions of years ahead of us, were true, then what could we learn from them? The belief that we can benefit from an advanced SC reveals a number of assumptions. First, that science is a universal activity and, second, that terms like 'advanced' and 'progress' are unproblematic. There is also the problem of knowing that what they do is actually science, when it has proved so difficult for scientists and philosophers on Earth to provide a satisfactory definition of science. At a basic level it could be said that science involves forming beliefs about the world, recognizing problems, formulating solutions and testing them by observations and experiment, and then applying them in practical contexts. It is not certain that all intelligent beings will be scientific. Dolphins are intelligent but not scientific, as intelligence must be put to use in some particular way.

Nicholas Rescher (1984) is one of the few philosophers to have described the problem of communicating with an advanced SC. He draws an important distinction between the '*content* of their doctrines (belief structures and theory complexes)' and 'the *aims and purposes* with which their doctrines are formed' (Rescher, 1984: 176). If we consider how the content of human enquiry has varied so much throughout human history – for example, from the speculations of the nature theorists of ancient Greece to the gulf between Newtonian and contemporary physics – then Rescher suggests that emphasis should be given to their aims and purposes: in short, to find out whether their projects are functionally equivalent to our scientific projects, seeking explanations, predictions, and control over nature. But when he considers the methods they might employ to express their science, Rescher notes enormous potential for diversity. They might have a different mathematics to ours, with anumerical approaches to quantity, stressing comparability rather than quantification. Such a mathematical system with a different geometry might prove to be more effective for application in an unstable environment: if, for example, they had structures like jellyfish and lived in the sea. They might not have developed digital thinking and their description of nature in mathematical terms might be very different to ours. Their orientation of science might be different; their natural science could well be oriented towards social categories. Rescher points out how nature is observed selectively. The ETs might very well ignore certain phenomena which are considered significant to us: for example, if their environment does not have magnetic poles, then observations of loadstones and electrical storms might have little scientific meaning.

Creatures that have evolved in very different natural circumstances to us will have very different interests which will influence the emphasis they place upon certain experiences. Creatures which have adapted to life underground, or beneath the sea, will never develop astronomy. Access to different natural facts will influence the topics which give rise to puzzlement and stimulate investigations. Different ways of experiencing nature will raise different questions about nature. A society of advanced ETs, with different structures and sensory receptors, might develop interests in parts of the environment that mean nothing to us.

Rescher argues that we must expect ETs to have different conceptual schemes, different ways of making sense of their experience, such that discourse with them might be impossible. The point made by Thomas Kuhn (1962) that there is no 'ideal scientific language' for the characterization of reality, is even more relevant, argues Rescher, when speaking of alien cultures. As a science develops, the concepts it uses evolve in the course of various discoveries and not only help to make these discoveries; they are in turn modified by them. The ancient Romans, says Rescher (1984: 181), did not have a different conception of quantum electrodynamics; they had no view at all about it. There were many steps from Cicero to Bohr; whole theories about the universe and matter had to be abandoned and others built up from scratch. With an alien civilization, argues

Rescher, there would be no common foothold and the problem of interpretation would be immense. Stressing the parochial nature of our science, Rescher (ibid.: 182) concludes that: 'Given intelligent beings with a physical and cognitive nature profoundly different from ours, one simply cannot assert with confidence what the natural science of such creatures would be like.'

Even with a common cultural heritage, for example, Western Christianity, there is room for widespread divergence and communicative failure. A shift to an extraterrestrial setting, argues Rescher, would amplify the diversity found on Earth. This raises serious problems with regard to proposals for communicating with 'scientifically advanced' ETs. The concept of 'advanced' is problematic: to be more advanced it would have to be a variety of our sort of science. Given differences in formulation, conceptualization and orientation, Rescher (ibid.: 188) sees the possibility of a similar science and technology as 'extremely remote'. To be more or less advanced presupposes a uniformity of objectives; we can say that they are more advanced than us if they do the same things that we do better or more fully. The concert pianist is better than the child beginner because each one is travelling in the same direction, aiming at similar objectives against a uniform standard. But, says Rescher (ibid.): 'we can scarcely say that Chinese cookery is more or less advanced than Roman, or Greek pottery than Renaissance glassware.' A science of alien beings, concludes Rescher, is bound to be different to ours, with a very small probability that they would have our scientific posture. There is no sense in speaking of them being more advanced or backward if they are not travelling on the same journey. 'It is thus farfetched to suppose that an alien civilization might be scientifically more advanced than we are' (ibid.: 174).

Rescher's argument clearly reduces the likelihood of communication with numerous civilizations, but does it rule out the possibility altogether? They may, for example, do things that we would regard as advanced science, such as harnessing the ocean's energy, or deriving energy from cold fusion, but do it in the context of a religious or aesthetic belief system which is completely beyond our comprehension. Yet our scientists may learn something from it. Moreover, even if we accept that being more or less advanced requires a uniformity of destination, this still leaves plenty of scope; they might not recognize our criteria for being either backward or advanced, playing the game by different rules, but they may be more advanced in some activities but not in others; they may be advanced in techniques of space flight, more advanced because their spaceships carry more weight, move faster, and so on. Dialogue with a society within which certain practices are recognized as more advanced than ours need not be ruled out; the more advanced in certain activities can develop skills in communicating with the backward, as teachers communicate with infants and children. Rescher assumes that we could not know if they were more advanced. But *they* might and they might also know what they had left behind – Kuhn-loss – as their science developed. Cicero might not be able to comprehend quantum electrodynamics, but neither could a child. Put them both in an environment where quantum electrodynamics was an accepted part of the prevailing belief structure and a good teacher should be able to instruct both of them. To be sure, the finest developments in, say, medical technology might not be appreciated by beings with a radically different physiology, but we could still appreciate measures they took to extend their life-spans. Moreover, a civilization composed of immortals, whose historical records indicate that mortality was defeated, by means of intelligent intervention, 10,000 generations ago, would clearly be recognized as being more advanced than us. They may have a very different concept of physical duration, and not see immortality in any way linked to medical science, but our medical science might well learn from it, just as contemporary science is capable of benefiting from developments external to it. We might not have the same physical environment or a similar physical structure or even the same cultural and scientific traditions, but we can still detect scope for our scientific advancement in their different systems and struggle to learn from them.

A restricted view of learning from advanced extraterrestrials was presented by Alfred Adler who maintained that:

> Something tangible might indeed be learned from an extraterrestrial civilization, but this would have to be within the realm of our present knowledge and capabilities, or else we could not comprehend its meaning. Being in this realm, it would be something we would sooner or later have discovered for ourselves, without outside help.
>
> (1980: 227)

According to Adler, we could learn very little from contact. But this is based on a very naïve view of learning. As long as the gap is not too wide – and we have set it so as to exclude Type III civilizations – it is possible for the very advanced to find or construct a set of meanings which would enable the very backward to learn, as we do when teaching children.

Consider a contemporary scientific problem: do tachyons exist? They have not been observed, but according to some scientists they are entities which never travel at less than the speed of light. Consequently tachyons, and theories of communication by means of tachyons, have been ruled out by many scientists as their existence would be contrary to Einstein's theory according to which nothing can exceed the speed of light. Yet tachyon theory is derived from an extension of Einstein's special relativity theory in 1905. Einstein's theory describes a mechanics of material particles which never travel faster than light. But the theory can be reformulated with particles of another very different kind – with no upper limit to their velocity, only a lower one. Thus if tachyons are supplied with energy their motion would actually decrease. If such particles exist, they could travel at optimum speeds of 100,000 times that of light, and an exchange of information across the entire universe could be completed in less than one hundred years. But they have not been observed and, more significantly, we have no idea as to how they could be transmitted or received.

Presumably, a Type II or III SC could answer the question whether tachyons exist or not by revealing their use of laws relating to them. If tachyons do exist and can be used, despite the negative results of our searches, then our physics and communications systems could, at least in theory, be transformed. For the existence of tachyons – even if we lacked the ability to employ a technology that could use them – would contradict some of the laws and principles which are central to our physics. However, it should be stressed that despite the tension between tachyon theory and prevailing views which reject the idea of faster-than-light travel, and the consequent transformation that knowledge concerning the existence of tachyons would have on on our physics, there are conceptual links from existing theory to tachyon theory, such that dialogue between those who know of their existence and we who are ignorant could take place.

A more advanced society might have a very advanced system of education and methods which, like our archaeologists, can seek understanding of civilizations of long ago. Moreover, if they can send and receive radio signals they would still remain within our communication threshold, and would consequently share some knowledge of the universal features of reality. They might have evolved beyond the use of radio but nevertheless retain it as an antiquarian interest, sending signals in a primitive language like enthusiasts in a post-industrial culture who employ spinning wheels and other pre-industrial techniques. These ET hobbyists might find it fascinating to communicate with primitives on Earth.

There are questions we could ask a superior civilization: how can the energy problem be solved? How did life get started? Can cancer be cured? How can we avoid destroying the world? There are questions that cannot be answered in terms of the science we know. How can we reconcile general relativity with quantum mechanics? Is the Big Bang theory of the origin of the universe a meaningful theory? Does the bulk of the universe consist of dark matter? Or is such a theory, like ether theory, a folly of our time? They could understand how we came to formulate theories about dark matter, the role it plays in our speculations about the expansion of the universe, and in reply explain either why we are on the wrong track – as in the case of ether theory – or how theories of dark matter are likely to develop into a major scientific breakthrough. Of course we would run the risk of posing questions where answers would totally mislead us. Consider, for example, a request for better systems of illumination in 1800: expected answers would be in terms of a technology linked to candles and oil lamps, not in terms of electricity. Or consider a sixteenth-century monk to whom an angel appears and writes $E = MC^2$. What use would this information be? It would be easier to work through the next three centuries of scientific history than to spend time pondering on the mystery of this symbol. In these cases there just is no chain of reasoning from one standpoint to the other. Science does not develop in a linear fashion; it draws on a wide range of developments in culture, art, religion, and many other disciplines. Thus if an angel were to present Michelangelo with a CD which explained nuclear technology he would not know what to make of it, as the progress of scientific learning takes place within a broad spectrum of discoveries and related techniques.

Yet in another sense the appeal to incommensurable frameworks is a false objection: the historians of an advanced civilization could be aware of dramatic conceptual revolutions, Kuhn-loss and paradigm changes, and acknowledge these when dealing with primitives. A twenty-first century physicist, with a sound history of science, would not deliver $E = MC^2$ for a sixteenth-century monk but would attempt to employ concepts that are known to have been familiar to sixteenth-century scholars. This might involve a complicated process of engaging in discourse about angels, but the more advanced of the two would be aware of the missing links in the chain and would endeavour to find some common ground. A historian of science with a good grounding in modern physics could very well communicate with Newton and no doubt help him solve some of his scientific problems.

Older civilizations may give us knowledge of the past. Imagine what our astronomers would make of photographs of the galaxy which were taken 5 billion years ago! It may be that much of what they do lies beyond our comprehension. Carl Sagan (1973: 366–7) suggests that: 'We might be like the inhabitants of the valleys of New Guinea who may communicate by runner or drum, but who are ignorant of the vast international radio and cable traffic passing over, around and through them.' But this does not prevent us from communicating with them. We may not want to, and advanced ETs may not want to communicate with us, but this would seem to be unlikely. Many of our anthropologists and historians would love to communicate with Stone Age people and ancient civilizations, and are prepared to go to great lengths in order to comprehend and translate the messages they have left behind. But this may be the limit of the interest of advanced ETs in us. If they are so advanced, and as the speculation goes, in contact with numerous SCs across the universe, they would not find us intellectually stimulating. For them the activities of the inhabitants of Earth would not be of major importance, and their interest would be limited to that of a Department of Anthropology.

Even with the best intentions there would very likely be a communication horizon, as Sagan observes (ibid.: 366). We might wish to communicate with our Stone Age ancestors, but not with our protozoan or bacterial forebears. The anthropologist may study old societies and wish to communicate with them, but not as far back as micro-organisms. On these terms then Kardaschev–Dyson SC Type IIs might be within our communicative horizon, but Type IIIs are unlikely to exist within our threshold and the best we could do is discover them, observe them and possibly admire them.

Yet the very knowledge of the existence of advanced civilizations would resolve certain questions and stimulate the search for solutions to others. This knowledge, for example, would raise profound questions about our destiny. In about 7 billion years our Sun will enter its red giant phase. Is this the end or can cultures survive the death of their stars? Is there a galactic network? How prevalent is life in the universe? How did life get started? What is the typical life-span of a technological civilization? Is there intergalactic space travel? If contact is made with ETI from within our galaxy we could assume that since at least two independent intelligent cultures have evolved here, there are likely to be others in many galaxies, which would suggest that life is widespread. We might also infer that they had learnt to survive technology, although there is always the possibility that they are sending a 'goodbye-don't-make-our-mistakes' warning from a dying civilization.

If contact is made with more advanced civilizations it may well be with robots who have replaced human intelligence, suggests Paul Davies (1995). This could, argues Davies, also have a profound effect on us, as it would reveal that mankind does not represent the pinnacle of evolutionary advance and this knowledge could have a demoralizing effect on people. On the other hand, the fact that knowledge can advance so much might prove exhilarating and inspiring: 'Either way,' says Davies (ibid.: 36): 'it is hard to see how the world's great religions could continue in anything like their present form should an alien message be received.' For if they had abandoned religion, this knowledge could damage our faith, but if they had retained a form of spirituality it would very likely convert many humans, argues Davies. Perhaps this forecast is too severe: most of the world's religions are very adaptive and can accept new concepts. Christianity has survived Copernicanism, the collapse of Aristotelianism, Darwinian theory and Marxism. Three centuries of scientific materialism have not eliminated the need for a spiritual dimension of experience and it is unlikely that knowledge concerning an extraterrestrial belief system would have a devastating effect. Moreover, religious fundamentalists would not have to adapt their dogmas; they are quite experienced in the rebuttal of scientific facts that appear to conflict with their beliefs, and could continue to reject whatever evidence is produced, as they have done with Darwinian theory or any other scientific position which they find unacceptable.

In an optimistic survey of the benefits of contact Steven J. Dick (1998: 253) maintains that the Judaeo-Christian as well as Buddhist, Hindu and Chinese religions should have no problem in adapting to the realities of ETI. However, he notes, 'No true astrotheology was developed in the twentieth century in the sense that new theological principles were created, or existing ones formally modified, to embrace other moral agents in the Universe.'

Many SETI enthusiasts claim that knowledge of the existence of an advanced technological society would demonstrate that intelligence can survive advanced technology without global war. But this argument may be naïve, as it assumes that our political experience will be inevitably reproduced with a similar technology. Other political controls may have been developed in which technology is not geared to war and high consumption, and would not pose the threat of ecological destruction. Nevertheless, it might be argued that we could still learn something from their alternative systems of political control. This has been disputed. According to Puccetti (1968), it is unlikely that we could learn how to solve our problems and survive technology from observations of intelligent extraterrestrials. Little practical information could be gained from reports of their political structures, as other cultures and their histories are never applicable to our own difficulties, he argues. 'This should be more obvious where the historical knowledge concerns a biologically distinct race or another planetary system' (ibid.: 106). This may be too pessimistic. We are capable of learning from different political structures and different cultures; most of the world's great religions and ethical systems are derived from the histories of people we have little in common with, and many of these are deities or extraterrestrials of one form or another.

SETI researchers have predicted that contact with a more advanced species would be beneficial. Optimistic SETI researchers speak of the benefits to be derived from contact with superior intelligent extraterrestrials. Frank Drake (Drake and Sobel, 1993: xiii) maintains that contact with a more advanced civilization would provide 'us with a glimpse of what Earth's future could be'. According to Michael Michaud (1990), contact could initiate a 'knowledge revolution'. Even the mere detection of ETI would bring new knowledge about the evolution of life and intelligence. If they sent a message describing their biological make-up it would certainly have a profound effect on our theories regarding the origins of life, especially if the message revealed an independent line of evolution. Even if we could not interpret their signals the fact of their existence would still tell us much about their technology. If we could interpret them that would certainly have a profound impact. They may introduce new aesthetic forms and new branches of science. Michaud compares the potential contact between ETI and Earth with the knowledge transmitted to medieval Europe from ancient Greece; it could be a new synthesis of knowledge, he says, heralding a great leap forward, as in the Renaissance. It would involve the end of cultural isolation: if there is a galactic network then we would not be the first civilization to make contact, and if contact exists it is likely to have been going on for some time. Moreover, if there really is a communication network then once we are connected to it we could learn of the history of the galaxy, about early civilizations, alternative cultures, arts, political systems and means of survival.

This optimism may turn out to have been ill-founded. Difference often breeds misunderstanding. They might not wish to share their knowledge, tell us how to make more powerful weapons or how to plunder the natural resources of the galaxy. Confrontation with superior knowledge could destroy the morale of scientists and undermine support for research. Communication itself could amount to a form of mental colonization if extraterrestrials have superior minds combined with manipulative tendencies. The very idea of a superior intelligence is hard to accept, especially for the human male who has problems accepting that other terrestrial animals have intelligence. Those on Earth who had access to information from a more advanced technology may reveal strong tendencies to monopolize this knowledge. The temptation to patent and then monopolize would be very strong among those with access to information from advanced ETs. This would be highly probable if the information was relevant to military or commercial interests. One obvious drawback to ETI communication is the risk of exploitation by cults and commercial interests. It might also be noted how leaders of UFO cults monopolize and then manipulate the alleged information from ETI in their own interests. There are already very powerful forces on Earth with a determination to control information. Knowledge of an advanced ETI technology could very well fall into the hands of these forces.

Societies on Earth have long histories of contact with superior beings, such as saints and reformers, even gods, without lasting success. What grounds do we have for assuming that a message from Tau Ceti would improve matters? Moreover, our history shows that the technologically 'inferior' do not benefit from contact with more advanced societies. If they are not exterminated or enslaved, they frequently fall victim to various pathologies, alcoholism, drug abuse and high suicide rates, which have been attributed to the influence of Western technological culture on various native cultures. Captain Cook's sailors introduced Christianity, syphilis and leprosy when they discovered Hawaii. In general the history of European exploration and discovery led to the spread of disease and cultural disintegration, and too often slavery.

Advanced ETs might not look upon us from a benevolent standpoint. There is always the chance that radio silence may conceal an impending invasion. They may come armed and we will either be annihilated or enslaved, or they will come in the guise of friendship and take us over by stealth by appearing to help us while making us dependent upon them. They might come and exploit us like pets, raw materials or food products. There is no reason to suppose that a superintelligence would develop superlevels of compassion and empathy.

Yet it is very unlikely that ETs would take us away as slaves or food products or steal our mineral resources; the vast distances would render such an outlay of energy uneconomical. But maybe they would have no choice but to come and would be forced to migrate as their star approached extinction. Yet if that were the case they would have had time to undertake reconnaissance and would most likely prefer to move to a habitable but uninhabited world. On the other hand, if they are so advanced and morally unscrupulous and Earth appeals to them, then our resistance would be negligible.

Other than enforced migration or invasion for our natural resources, the main commodity they would seek is information, which costs little to transport. In fact, information would be the most likely commodity they would look for. But this would still leave open the threats associated with a culture clash. On the other hand, SETI researchers will argue that the culture clash would not have to be damaging. We have all experienced contact with minds much greater than ours – in the great literature and science of the world – without negative effects, and such contact is frequently inspirational.

Perhaps contact would not have such a profound effect on the scientific community, as the fundamental shift in thinking about the possibility of ETI has already taken place throughout the twentieth century. During the latter half of the twentieth century there was a gradual awareness that the processes which gave rise to life and continue to support it, could be widespread. More generally, belief in the existence of ETI is a strong feature within our culture, even when it is associated with discredited beliefs in UFO visitors. So rather than seeing the discovery of ETI as heralding a scientific revolution, it might be seen as an important stage in a series of changes to our view of life in the universe.

Contact would certainly inspire discussion of the age-old philosophical problems of where do we come from, are we unique, and what does it mean to be a human being? Moreover, if there is a galactic community, possessing knowledge of how to survive technology and prevent cultural destruction, then membership of this community would likely produce a pride in identification with a supersociety; far from a loss of morale, there would be an intense pride in membership. If contact is by radio, the effects of culture shock would be reduced because of the lengthy communication time involved. It will take time to verify that the contact is of artificial origin, then more time to decipher the message and eventually achieve consensus regarding the content of the version that will be published. The research group authorized to decipher the message would probably sit on it for years until public interest has long evaporated. The initial news will be impressive but it will be hard for it to overshadow the hype that currently accompanies trivial events. A royal marriage, a soccer star's divorce or an adulterous liaison between characters in a soap opera, will easily push news of alien contact to the back pages.

Radio communication, then, would not be like a swift invasion or sudden confrontation with an alien culture. In the latter case, a well-authenticated physical presence might well produce an effect similar to the notorious *War of the Worlds* broadcast. This is probably unlikely, as knowledge of ETI would precede any physical appearance and years of popular science and SF have prepared the public mind for contact. Consequently, the knowledge that we are not unique is unlikely to have any destabilizing influence.

8

CONCLUSION

Accounts of supercivilizations and contact with them have a long history in science fiction and have undoubtedly influenced progressive scientists who believe that mankind might one day benefit from such contact. But, like religious accounts of divine beings with superhuman knowledge and power, they ultimately refer back to terrestrial concerns. In fact, most accounts of SCs are manufactured by emphasizing one aspect of the technological developments achieved on Earth over others. Thus, for example, Kardaschev's version of an advanced civilization emphasizes energy consumption and Dyson's emphasizes demographic forces, the effect of which are demonstrably significant for terrestrial life. The entire SETI endeavour reflects interest in technological developments on Earth. Thus advances in radio astronomy since the Second World War prepared the ground for NASA and the SETI community's microwave searches, while more recent developments in laser technology have encouraged expectations of an optical signal. There are other versions of extraterrestrial civilizations which emphasize aesthetic or spiritual qualities and consequently reject technology and the very idea of colonization and, like some of Earth's societies, prefer to look inwards for salvation. One version of an advanced extraterrestrial civilization at work is Ferris and Bracewell's idea of a network of self-replicating communicating probes. This clearly emphasizes current terrestrial interest in robots, artificial intelligence and the replication of cognitive skills such as knowledge and prediction. The information technology culture is also an influential factor in Scheffer's theory of teleportation by information transfer. Even the wildest theories concerning UFO visits and government cover-ups are reflections of serious terrestrial concerns. They might be dismissed as the ravings of paranoid crackpots but in a deeper sense they give expression to a sound healthy scepticism in the face of an unholy and secretive alliance between governments, the military and the scientists who serve them.

These different models of extraterrestrial civilizations have clear implications for any search strategy. Although emphasizing a search by means of infrared astronomy, both Kardaschev and Dyson's models would entail a passive search strategy which involves little more than waiting to see if ETI can be discovered by means of existing observational technology. Tipler's appeal to colonization would lead to a search for spaceships, not microwaves. Likewise, Zuckerman's account of obligatory mass migrations would involve a search for expeditionary ventures, and Stephenson's proposal regarding travelling biospheres would require a search of the outer regions of the solar system as a likely zone for the detection of extraterrestrial vessels. But if the conclusions of Hart and Tipler are correct, then SETI is a waste of time. Spiritual and aesthetic models are notorious for their appeal to subjective psychological experience, while the postulation of Bracewell probes entails a passive wait-until-they-come attitude. The Peer Hypothesis, developed by Paul Birch, would involve a search with military expectations, while most versions of the Zoo Hypothesis involve an eavesdropping approach. The practical outcome of radio searches will involve requests for bigger, better and more widely based searches, preferably conducted in space.

Yet the one thing that can be predicted is that most of our predictions about contact with an advanced civilization are likely to be wrong. So far no society on Earth has correctly predicted its future: no ancient Greek, Egyptian or Babylonian predicted television or its effect on human culture. Most SC models neglect the reality of the changing direction of cultural progress. Religion on Earth was once the foremost vehicle for social progress and that would have been the area to look to for predictions regarding future cultural developments. But religion gave way to science, although it may even return again to its former exalted status. New, as yet unheard-of problems may shape our conception of progress and notion of progressive forces. Many exponents of SETI research assume that science and technology will be valued for their own sake and be seen in the forefront of progress. But maybe our need for survival may lead to a replacement of those values.

Notwithstanding the philosophical and methodological problems that have been raised against SETI, there are several compelling reasons for continuing. Proposals for SETI research are fundamentally committed to the empiricist methodology. SETI is also essentially interdisciplinary. No subject shows itself as capable of bringing together the fragmented elements of contemporary science. SETI embraces branches of astronomy, physics, biology, psychology, computer science, cognitive science, linguistics, information theory, natural and social history, geology, zoology, philosophy, theology, sociology and political science. SETI is also a discipline that is ripe for creative thinking, and a broader paradigm of scientific inquiry may emerge from it, drawing on the insights of poets, theologians and philosophers.

Apart from its unifying contribution to our understanding of scientific research, SETI and space exploration have lent support to environmental ethics: the view from space of Earth as a delicate and fragile world has motivated people to think of its ecological protection. Few Westerners had a concept of ecology until astronaut Frank Borman referred to 'Spaceship Earth'. The opportunity to look back on Earth as a fragile oasis of life may turn out to be as significant as the Copernican revolution.

CONCLUSION

The exploration of space, the search for life and extraterrestrial intelligence, will not merely add to the weight of scientific facts; it will play a fundamental role in the transformation of science, its goals and methods, as well as a transformation of our relationship to the world around us. For scientific research is not merely about the accumulation of facts, it is also bound up with the transformation of facts and theories and inevitably with the transformation of human expectations. Space research in general will undoubtedly bring benefits, many of which will be unpredictable, with inevitable spin-offs affecting other branches of life, and some of them will introduce new problems and terrors which future generations will have to live with.

There is a need to recognize the importance of curiosity-driven research, not merely freedom for a few researchers to gather esoteric data, but in the service of a curiosity that is integral to our culture and the way it has evolved over the past two millennia. For a fundamental reason for SETI research is natural curiosity. Despite non-curious governments, curiosity has a way of triumphing in many guises, one of which involves persuading potential investors of profitable side-effects. The discovery of new continents in the sixteenth and seventeenth centuries, the opening up of new frontiers in the nineteenth century and the cultural expansion of the twentieth century, have all fostered a tradition of exploration that is still strong enough to excite and capture the imagination as did the first lunar landings over a quarter of a century ago. The search for life in the universe is, perhaps, one of the last great frontiers of science. The information technology culture may be an Earthbound phenomenon, and many of SETI's scientists point out that electronic information, not voyages, is what they have in mind. But if a signal is detected, an electronic exchange will be considered insufficient and we can expect renewed inspiration in the direction of manned space flight.

Recognition of the precarious status of life on Earth may herald a renewed interest in colonization. Repeated warnings of asteroid collisions with the Earth and similar global catastrophes which, in the past, have decimated life on Earth, may drive home the point that the human race has too many eggs in one fragile basket. For if we are alone it might be argued that we have a duty to export life.

The drive to discover and contact other intelligences will inevitably continue. In the belief that governments know of them but will not tell, in NASA's and the SETI community's radio searches and other searches throughout the world, and also among the UFO watchers, there is one common theme, a common spirit. They participate in an age-old conscious decision to act on their daydreams about other worlds, expressing the hope of contacting other souls across the universe and combating cosmic loneliness. Until very recently, arguments for and against the existence of extraterrestrial life have been confined to speculation and science fiction, but we now have the technology to conduct an empirical investigation. If life is common, then evidence might be found on Mars or Europa within the next twenty years. If extraterrestrial intelligence is found and contact is made, it will be truly important. If we do make contact our children will be astonished to discover that we made so little effort to do so, and then they will laugh at those who denied any possibility of contact. But if, after a massive search, we fail, that too will be important, as it will convince many of us that if this is all there is, we should do our best to protect it.

BIBLIOGRAPHY

Adamski, G. (1953) *Flying Saucers Have Landed*, New York: Dell.
—— (1955) *Inside the Spaceships*, London: Penguin.
Adler, A. (1980) 'Behold the stars', in D. Goldsmith (ed.) *The Quest for Extraterrestrial Life*, Mill Valley, CA: University Science Books: 224–7.
—— (1999) 'Beyond the stars', *New Scientist*, 18 December: 4.
Andrews, G.C. (1987) *Extra-Terrestrials Among Us*, St. Paul, Minnesota: Llewellyn.
Annis, J. (1999) 'An astrophysical explanation for the "Great Silence" ', *Journal of the British Interplanetary Society*, January, 52: 19–22.
Anon (1994) 'This Week', *New Scientist*, 29 January: 7.
Ball, J.A. (1980) 'The zoo hypothesis', in D. Goldsmith (ed.) *The Quest for Extraterrestrial Life*, Mill Valley, CA: University Science Books: 241–5.
Barclay, D. (1994) 'Towards a full explanation', in D. Barclay and T.-M. Barclay, *UFOs: The Final Answer? Ufology for the 21st Century*, London: Blandford: 172–89.
Bauval, R. and Hancock, G. (1996) 'Could the Sphinx of Mars be proof of an ancient civilization as great as our own?', *Daily Mail*, 17 August: 40–2.
Benford, G. (1990) 'Alien technology', in B. Bova and B. Preiss (eds) *First Contact: The Search for Extraterrestrial Intelligence*, London: Headline: 212–28.
Bennett, C.H., Brassard, G., Crepeau, C., Jozsa, R., Peres, A. and Wooters, W.K. (1993) 'Teleporting an unknown quantum state via dual classical and Einstein-Podolsky-Rosen channels', *Physics Review Letters*, 70: 1895.
Birch, P. (1990) 'The peer hypothesis', *Journal of the British Interplanetary Society*, March: 1–6.
Blackmore, S. (1994a) *Horizon*, BBC2, 28 November.
—— (1994b) 'Alien abduction', *New Scientist*, 19 November: 29–31.
Blair, D.G. (1992) *Monthly Notices of the Royal Astronomical Society*, vol. 251: 105.
Blue Book Information Office (1968) *Project Blue Book*, WD Headquarters US Air Force, (SAF-OIPB), The Pentagon, Washington, DC.
Blum, H. (1990) *Out There*, New York: Simon & Schuster.
Bord, J. and Bord, C. (1992) *Life Beyond Planet Earth: Man's Contact with Space People*, London: Grafton.
Bova, B. (1990) 'Surveying the cosmos', in B. Bova and B. Preiss (eds) *First Contact*, London: Headline: 5–26.
Bracewell, R.N. (1974) 'Interstellar probes', in C. Ponnampereuma and A.G.W. Cameron (eds) *Interstellar Communication: Scientific Perspectives*, Boston: Houghton Mifflin Co.: 102–16.

—— (1976) *The Galactic Club: Intelligent Life in Outer Space*, San Francisco: San Francisco Book Co., Inc.

Brahic, A. (1994) 'The primordial nebula', in J. Audouz and G. Israel (eds) *The Cambridge Atlas of Astronomy*, 3rd edition, Cambridge: Cambridge University Press: 60–1.

Breuer, R. (1982) *Contact with the Stars: The Search for Extraterrestrial Life*, trans. by C. Payne-Gaposchkin and M. Lowery, Oxford: W.H. Freeman & Co. Ltd.

Brin, D. (1990) 'Mystery of the Great Silence', in B. Bova and B. Preiss (eds) *First Contact*, London: Headline: 154–83.

Campbell, B. (1990) 'A place to live: exsolar planets', in B. Bova and B. Preiss (eds) *First Contact*, London: Headline: 129–49.

Chown, M. (1995a) 'The ultimate message in a bottle', *New Scientist*, 18 February: 7.

—— (1995b) 'Stargazers amazed by "crazy" planet', *New Scientist*, 14 October: 18.

—— (1996) 'Planes, trains and wormholes', *New Scientist*, 25 March: 29–33.

Clarke, A.C. (1968) *2001: A Space Odyssey*, London: Arrow.

Cocconi, G. and Morrison, P. (1959) 'Searching for interstellar communication', *Nature*, 184: 844–6.

Coghlan, A. (1999a) 'A taste for Martian living', *New Scientist*, 12 June: 14.

—— (1999b) 'Space sickness', *New Scientist*, 12 June: 15.

Condon, E.U. (1968) *Scientific Study of Unidentified Flying Objects*, New York: Bantam Books.

Cousins, W.F. (1972) *The Solar System*, London, cited in Crowe (1988: 553).

Crawford, I.A. (1995) 'Some thoughts on the implications of faster-than-light interstellar space travel', *Quarterly Journal of the Royal Astronomical Society*, 36: 205–18.

Crick, F. (1981) *Life Itself: Its Origin and Nature*, London: Macdonald.

Crick, F. and Orgel, L.E. (1980) 'Directed Panspermia', in D. Goldsmith (ed.) *The Quest for Extraterrestrial Life*, Mill Valley, CA: University Science Books: 34–7.

Crosswell, K. (1991a) 'Trio of planets found orbiting nearby pulsar', *New Scientist*, 14 December:19.

—— (1991b) 'Pulsar's planetary exchange is real say astronomers', *New Scientist*, 21–28 December: 11.

—— (1992a) 'Puzzle of the pulsar planets', *New Scientist*, 18 July: 40–3.

—— (1992b) 'Why intelligent life needs giant planets', *New Scientist*, 24 October: 18.

Crowe, M.J. (1988) *The Extraterrestrial Life Debate: 1750–1900: The Idea of a Plurality of Worlds from Kant to Lowell*, Cambridge: Cambridge University Press.

Däniken, E. von (1969) *Chariots of the Gods?*, London: Souvenir Press.

—— (1970) *Return to the Stars*, London: Souvenir Press.

—— (1973) *The Gold of the Gods*, London: Souvenir Press.

—— (1974) *In Search of Ancient Gods*, London: Souvenir Press.

Davies, J. (1995) 'Searching for alien Earth', *New Scientist*, 13 May: 24–5.

Davies, P. (1995) *Are We Alone?*, Harmondsworth: Penguin.

—— (1998) 'Survivors from Mars', *New Scientist*, 12 September: 24–9.

Dawkins, R. (1988) *The Blind Watchmaker*, London: Penguin.

Dayton, L. (1999) 'Tiny wonders', *New Scientist*, 27 March: 13.

Dick, S.J. (1996) *The Biological Universe*, Cambridge: Cambridge University Press.

—— (1998) *Life on Other Worlds: the 20th-Century Extraterrestrial Life Debate*, Cambridge: Cambridge University Press.

Drake, F. (1990) 'The Drake Equation: a reappraisal', in B. Bova and B. Preiss (eds) *First Contact*, London: Headline: 150–3.

BIBLIOGRAPHY

Drake, F. and Sobel, D. (1993) *Is Anyone Out There? The Scientific Search for Extraterrestrial Intelligence*, London: Souvenir Press.

Dyson, F.J. (1960) 'Search for artificial stellar sources of infrared radiation', *Science*, 131: 1667; reprinted in D. Goldsmith (ed.) (1980) *The Quest for Extraterrestrial Life*, Mill Valley, CA: University Science Books.

—— (1973) 'Discussion', in C. Sagan (ed.) *Communication with Extraterrestrial Intelligence: CETI*, Cambridge, MA: MIT Press: 188–9.

Ferris, T. (1992) *The Mind's Sky*, London: Bantam.

Feyerabend, P.K. (1975) *Against Method*, London: New Left Books.

Firsoff, V.A. (1963) *Life Beyond the Earth: A Study in Exobiology*, London: Hutchinson.

Freudenthal, H. (1960) *Lincos: Design of a Language for Cosmic Intercourse, Part I*, Amsterdam: North Holland Publishers.

Gold, T. (1960) 'Cosmic garbage', *Airforce and Space Digest*, May: 65.

Goldsmith, D. (ed.) (1980) *The Quest for Extraterrestrial Life: A Book of Readings*, Mill Valley, CA: University Science Books.

Good, T. (1991) *The UFO Report 1991*, London: Sidgwick and Jackson.

—— (1992) *The UFO Report 1992*, London: Sidgwick and Jackson.

—— (1993a) *The UFO Report 1993*, London: Sidgwick and Jackson.

—— (1993b) *Alien Update*, London: Arrow.

Gott, J.R. III (1993) 'Implication of the Copernican Principle for our future prospects', *Nature*, 363, 27 May: 315–19.

Gould, S.J. (1994) 'The evolution of life on Earth', *Scientific American*, October: 63–70.

Greer, S.M. (1992) *CSETI*, Ashville, NC: Greer.

Gribbin, J. (1991) 'Is anyone out there?', *New Scientist*, 25 May: 29–32.

—— (1994) *In the Beginning: The Birth of the Living Universe*, Harmondsworth: Penguin.

Hart, M.H. (1975) 'An explanation for the absence of extraterrestrials on Earth', *Quarterly Journal of the Royal Astronomical Society*, 16: 128–35.

—— (1980) 'N is very small', in M.D. Papagiannis (ed.) *Strategies for the Search for Life in the Universe*, Dordrecht: D. Reidel: 19–25.

Hecht, J. (1994) '"Molecule of life" is found in space', *New Scientist*, June: 4.

—— (1996) 'Space array could find alien worlds', *New Scientist*, 17 February: 10.

—— (1997) 'Titan awakes', *New Scientist*, 29 November: 28.

Heidmann, J. (1989) *Life in the Universe*, New York: McGraw-Hill Inc.

—— (1995) *Extraterrestrial Intelligence*, Cambridge: Cambridge University Press.

—— (1997) *Extraterrestrial Intelligence*, 2nd edition, Cambridge: Cambridge University Press.

Heine, M. (1991) 'The erotic ontology of cyberspace', in M. Benedikt (ed.) *Cyberspace: The First Steps*, Cambridge, MA: MIT Press: 59–80.

Henbest, N. (1992a) 'SETI: the search continues', *New Scientist*, 10 October: 12–13.

—— (1992b) 'When will Earthlings see the light?', *New Scientist*, 12 December: 48.

Holmes, B. (1996) 'Death knell for Martian life', *New Scientist*, 21–28 December: 4.

Horgan, J. (1994) 'Beyond Neptune,' *Scientific American*, October: 13–14.

Horowitz, P. (2000) 'Optical SETI joins the search', *The Planetary Report*, XX, 2, March–April: 8–12.

Horowitz, P. and Alschuter, W.R. (1990) 'The Harvard SETI search', in B. Bova and B. Preiss (eds) *First Contact*, London: Headline: 261–70.

Horrowitz, N.H. (1977) 'The search for life on Mars', *Scientific American*, 237, November: 52–61.

Hoyle, F. and Crick, F. (1981) *Life Itself*, New York: Simon & Schuster.

Hoyle, F. and Wickramasinghe, C. (1977) 'Does epidemic disease come from space?', *New Scientist*, 76, 17 November: 402.

—— (1978) *Life Cloud: The Origin of Life in the Universe*, London: J.M. Dent.

Hughes, D. (1992) 'When planets boldly grow', *New Scientist*, 12 December: 29–33.

Hynek, J.A. (1972) *The UFO Experience: A Scientific Inquiry*, Chicago: Henry Regnery.

—— (1977) *The Hynek UFO Report*, London: Sphere.

Impey, C.D. (1995) 'The search for life in the Universe', *Vistas in Astronomy*, 39: 553–71.

Jakosky, B. (1996) 'Warm havens for life on Mars', *New Scientist*, 4 May: 39–42.

—— (1998) *The Search for Life on Other Planets*, Cambridge: Cambridge University Press.

Johansson, B. (1998) Letter, *New Scientist*, 14 February: 53.

Kant, I. (1949) *Critique of Practical Reason*, trans. L. White Beck, Chicago: Chicago University Press.

—— (1956) *Critique of Pure Reason*, trans. J.M.D. Meiklejohn, London: Macmillan.

—— (1981) *Universal Natural History and Theory of the Heavens*, trans. S.L. Jaki, Edinburgh: Edinburgh University Press.

Kardaschev, N.S. (1964) 'Transmission of information by extraterrestrial civilizations', *Astrononicheskii*, zh. 41: 282, English translation in *Soviet Astronomy*, A.J. 8: 217.

Katz, I. (1992) 'Scientist confesses planet find fails to add up', *The Guardian*, 16 January.

Kepler, J. (1967) *Kepler's Somnium: The Dream or Posthumous Work on Lunar Astronomy*, trans. E. Rosen, Madison, WI: University of Wisconsin Press.

Kiernan, V. (1994) 'Antarctic rock could give clue to life on Mars', *New Scientist*, 5 March: 12.

—— (1995) 'The day the aliens failed to land', *New Scientist*, 5 August: 10.

—— (1997) 'NASA told how to avoid Mars attack', *New Scientist*, 22 March: 6.

Kiernan, V., Hecht, J., Cohen, P. and Concar, D. (1996) 'Did Martians land in Antarctica?', *New Scientist*, 17 August: 4–5.

Kinder, G. (1988) *Light Years*, London: Penguin.

Kleiner, K. (1997) 'Private enterprise boldly goes', *New Scientist*, 20 September: 18.

Knight, J. (2000) 'Buried evidence', *New Scientist*, 18 March: 11.

Krauss, L. (1996) 'Illogical Captain', *New Scientist*, 20 April: 24–9.

Kuhn, T. (1957) *The Copernican Revolution*, Chicago: Chicago University Press.

—— (1962) *The Structure of Scientific Revolutions*, Chicago: Chicago University Press.

Kuiper, T.B.H. and Morris, M. (1977) 'Searching for extraterrestrial civilizations', *Science*, 196: 616–21.

Lamb, D. (1991) *Discovery, Creativity and Problem Solving*, Aldershot: Ashgate.

—— (1994) 'Crop patterns and the greening of Ufology', *Explorations in Knowledge*, XI, 2: 12–46.

—— (1997) 'Communication with extraterrestrial intelligence: SETI and scientific methodology', in D. Ginev and R.S. Cohen (eds) *Issues and Images in the Philosophy of Science*, Dordrecht: Kluwer.

Learned, J.G., Pakvasa, S., Simmons, W.A. and Tata, X. (1994) 'Timing data communication with neutrinos: a new approach to SETI', *Quarterly Journal of the Royal Astronomical Society*, 35: 321–9.

Lebedev, N. (1991) 'The Soviet scene 1990', in T. Good (ed.) *The UFO Report 1992* (1992) London: Sidgwick and Jackson: 64–79.

Lovelock, J.E. (1979) *The Gaia Hypothesis*, Oxford: Oxford University Press.

Lowell, P. (1906) *Mars and its Canals*, New York: Macmillan.

—— (1909) *Mars as the Abode of Life*, New York: Macmillan.

MacArthur, R.H. and Wilson, E.O. (1967) *The Theory of Island Biogeography*, Princeton, NJ: Princeton University Press.

McDonald, G. (1998a) 'The stuff of life: must it be carbon-based?', *The Planetary Report*, March–April: 16–17.

—— (1998b) 'The Mars rock: some of its chemistry is from Earth', *The Planetary Report*, May–June: 9–10.

McDonough, T.R. (1987) *The Search for Extraterrestrial Intelligence*, London: John Wiley and Sons.

McGowan, R.A. and Ordway, F.I. III (1966) *Intelligence in the Universe*, New Jersey: Prentice-Hall.

McInnis, D. (1999) 'The next generation and the next ...', *New Scientist*, 25 December: 42–3.

McMullan, E. (1971) Review of R. Puccetti, *Persons: A Study of Possible Moral Agents in the Universe, Icarus*, 14: 291–4.

McNally, D. (ed.) (1994) *The Vanishing Universe: Adverse Environmental Impacts on Astronomy*, Cambridge: Cambridge University Press.

Marochnik, L.S. and Mukhin, M. (1988) 'Belt of life in the Galaxy', in G. Marx (ed.) *Bioastronomy: The Next Steps*, Dordrecht: Kluwer: 49–59.

Matthews, R. (2000) 'Star trecking', *New Scientist*, 15 April: 12.

Mayor, M. and Queloz, D. (1995) 'A Jupiter-mass companion to a solar-type star', *Nature*, 378: 355–9.

Meaden, G.T. (1989) *The Circles Effect and Its Mysteries*, Bradford on Avon: Artetech Publishing Co.

—— (1990) 'Crop circles and the plasma vortex', in R. Noyes (ed.) *The Crop Circle Enigma*, Bath: Gateway Books: 76–98.

—— (1991) 'Circles from the sky: a new topic in atmospheric research'. in G.T. Meaden (ed.) *Circles from the Sky*, London: Souvenir Press: 11–52.

Michaud, M. (1990) 'A unique moment in human history', in B. Bova and B. Preiss (eds) *First Contact*, London: Headline: 307–29.

Mobberley, M. (1994) Report of an address by Dr Tony Cook, 'Destination Mars', at the Department of Physics and Astronomy, University of Leicester, April 1994, *Journal of the British Astronomical Association*, 104, 6: 319–20.

Muir, H. (2000) 'The drifters', *New Scientist*, 1 April: 14.

Mullin, J. (1995) 'Keep watching the skies', *The Guardian*, 31 October: 2–3.

Mullins, J. (1995) 'The mine on the Moon', *New Scientist*, 18 November: 32–5.

Munévar, G. (1981) *Radical Knowledge*, Aldershot: Avebury.

—— (1989) 'Extraterrestrial and human science', *Explorations in Knowledge*, VI, 2: 1–8.

—— (1998) *Evolution and the Naked Truth*, Aldershot: Avebury.

Munévar, G., Preston, J. and Lamb, D. (2000) *The Worst Enemy of Science*, Oxford: Oxford University Press.

Nadis, S. (1994) 'Mars: the final frontier', *New Scientist*, 5 February: 28–31.

Oberg, J. (1995) 'Terraforming', in B. Zuckerman and M. Hart (eds) *Extraterrestrials: Where Are They?*, 2nd edition, Cambridge: Cambridge University Press: 86–91.

O'Hanlon, L. (1995) 'Low life would be at home on Mars', *New Scientist*, 28 October: 19.

Oliver, B.M. and Billingham, J. (eds) (1973) *Project Cyclops: A Design Study of a System for Detecting Extraterrestrial Life*, NASA Report CR-114445, Moffet Field, CA: NASA/Ames Research Center.

O'Neill, G.K. (1975) 'Space colonies and energy supplies to the Earth', *Science*, 190: 943–7.

Pain, S. (1998) 'The intraterrestrials', *New Scientist*, 7 March: 28–32.
Papagiannis, M.D. (1978) 'Are we all alone, or could they be in the Asteroid Belt?', *Quarterly Journal of the Royal Astronomical Society*, 19: 277–81.
—— (1988) 'Regional jurisdiction in our Galaxy: a possible explanation for the absence of extraterrestrial signals', in G. Marx (ed.) *Bioastronomy: The Next Steps*, Dordrecht: Kluwer: 281–5.
Pearce, F. (1994) 'Yes, we have no dead extraterrestrials', *New Scientist*, 1 October: 4.
Piaget, J. (1972) *Psychology of Intelligence*, Littlefield: Adams and Co.
Ponnampereuma, C. and Cameron, A.G.W. (eds) (1974) *Interstellar Communication: Scientific Perspectives*, Boston: Houghton Mifflin Co.
Pool, R. (1996) 'Time capsule older than Earth', *New Scientist*, 20 April: 17.
Puccetti, R. (1968) *Persons: A Study of Possible Moral Agents in the Universe*, London: Macmillan.
Purcell, E. (1980) 'Radioastronomy and communication through space', in D. Goldsmith (ed.) *The Quest for Extraterrestrial Life*, Mill Valley, CA: University Science Books: 188–96.
Randles, J. (1988) *Abduction*, London: Robert Hale.
Randles, J. and Fuller, P. (1990) *Crop Circles: A Mystery Solved*, London: Robert Hale.
—— (1991) 'Crop circles: a scientific answer to the UFO mystery', in G.T. Meadon (ed.) *Circles from the Sky*, London: Souvenir Press: 92–121.
Rardin, T.P. (1982) 'A rational approach to the UFO problem', in P. Grim (ed.) *Philosophy of Science and the Occult*, New York: SUNY: 256–66.
Rather, J.D.G. (1988) 'Optical lasers for CETI', in G. Marx (ed.) *Bioastronomy: The Next Steps*, Dordrecht: Kluwer: 381–8.
Reichhardt, T. (1996) 'Such stuff as worlds are made of', *New Scientist*, 13 April: 19.
Rescher, N. (1984) *The Limits of Science*, Los Angeles: University of California Press.
Rhodes, C. (1902) *The Last Will and Testament of Cecil J. Rhodes*, ed. W.T. Stead, London: Review of Reviews Office.
Ruzmaikina, T.V., (1988) 'Distribution of planetary systems', in G. Marx (ed.) *Bioastronomy: The Next Steps*, Dordrecht: Kluwer: 41–7.
Sagan, C. (1970) *Planetary Exploration*, Eugene, OR: Oregon State System of Higher Education.
—— (1972) 'UFO: The extraterrestrial and other hypotheses', in C. Sagan and D. Menzel (eds) *UFO: A Scientific Debate*, Ithaca, NY: Cornell University Press: 265–75.
—— (1973) *Communication with Extraterrestrial Intelligence: CETI*, Cambridge, MA: MIT Press.
—— (1980) *Cosmos*, New York: Doubleday.
—— (1983) 'The solipsistic approach to extraterrestrial intelligence', *Quarterly Journal of the Royal Astronomical Society*, 24: 113–21.
Sardar, Z. and Ravetz, J.R. (eds) (1996) *Cyberfutures: Culture and Politics on the Information Superhighway*, London: Pluto Press.
Scargle, J.D. (1988) 'Planetary detection techniques: an overview', in G. Marx (ed.) *Bioastronomy: The Next Steps*, Dordrecht: Kluwer: 79–82.
Scheffer, L.K. (1994) 'Machine intelligence, the cost of interstellar travel and Fermi's Paradox', *Quarterly Journal of the Royal Astronomical Society*, 35: 157–75.
Schlemmer, P.V. and Jenkins, P. (1993) *The Only Planet of Choice – Essential Briefings from Deep Space*, Bath: Gateway Books.
Schwartz, R.N. and Townes, C.H. (1961) 'Interstellar and interplanetary communication by optical masers', *Nature*, 190: 205–8.

BIBLIOGRAPHY

Searle, J. (1987) 'Minds and brains without programs', in C. Blackmore and S. Greenfield (eds) *Mindwaves*, Oxford: Blackwell Publishers: 208–53.

Seife, C. (1997) 'Exotic planet is gone with the wind', *New Scientist*, 20–27 December: 12.

—— (1998a) 'Money for old rock', *New Scientist*, 8 August: 20–1.

—— (1998b), 'Brave new worlds', *New Scientist*, 11 July: 23.

Shapere, D. (1991) 'The universe of modern science and its philosophical exploration', in E. Agazzi and A. Cordero (eds) *Philosophy and the Origin and Evolution of the Universe*, Dordrecht: Kluwer: 87–202.

Simpson, G.G. (1964a) *This View of Life*, New York: Harcourt, Brace & World Inc.

—— (1964b) 'The non-prevalence of humanoids', *Science*, 143: 769–75.

Simpson, J.A. (ed.) (1994) *Preservation of Near-Earth Space for Future Generations*, Cambridge: Cambridge University Press.

Smith, A.E. (1989) *Mars: The Next Step*, Bristol: Adam Hilger.

Space Science Board of the USA (1962) *A Review of Space Research*, National Academy of Science, National Research Council Publications, 1079, Chapter 9.

Stark, A.A. (1988) 'The Galactic centre: a nice place to visit but you wouldn't want to live there', in G. Marx (ed.) *Bioastronomy: The Next Steps*, Dordrecht: Kluwer: 291–3.

Stephenson, D.G. (1980) 'Extraterrestrial culture within the solar system', in D. Goldsmith (ed.) *The Quest for Extraterrestrial Life*, Mill Valley, CA: University Science Books: 246–9.

Stern, S.W.P. (1962) Review of Hans Freudenthal, *Lincos: Design of a Language for Cosmic Intercourse, Part I, The British Journal of the Philosophy of Science*, XII: 332–8.

Sullivan, W.T. III (1980) 'Radio leakage and eavesdropping', in M.D. Papagiannis (ed.) *Strategies for the Search for Life in the Universe*, Dordrecht: D. Reidel: 227–39.

Sweeney, J. (1995) 'Scanner of the alien nation', *Observer*, 13 August: 25.

Tarter, J. (1990) 'SETI: the farthest frontier', in A. Scott (ed.) *Frontiers of Science*, Oxford: Blackwell Publishers: 185–99.

—— (1995) 'One attempt to find where they are: NASA's High Resolution Microwave Survey', in B. Zuckerman and M. Hart (eds) *Extraterrestrials: Where Are They?*, 2nd edition, Cambridge: Cambridge University Press: 9–19.

Taylor, J. (1974) *Black Holes*, London: Collins.

Tipler, F.J. (1980) 'Extraterrestrial intelligent beings do not exist', *Quarterly Journal of the Royal Astronomical Society*, 21: 267–81.

—— (1995) *The Physics of Immortality*, London: Macmillan.

Veggeberg, S. (1993) 'Escape from biosphere 2', *New Scientist*, 25 September: 22–4.

Verschuur, G.L. (1980) 'A search for narrow bank 21-cm wavelength signals from ten nearby stars', in D. Goldsmith (ed.) *The Quest for Extraterrestrial Life*, Mill Valley, CA: University Science Books: 142–51.

Vidal-Madjar, A. (1994) 'The extraterrestrial life debate', *The Cambridge Atlas of Astronomy*, Cambridge: Cambridge University Press: 412–17.

Walker, G. (1996) 'Alien worlds boost search for life', *New Scientist*, 27 January: 17.

—— (1998) 'Did Hubble catch a glimpse of a nearby giant?', *New Scientist*, 31 January: 6.

Ward, M. (1997) 'Did Sally die in vain?', *New Scientist*, 17 May: 49.

Weston, A. (1988) 'Radio astronomy as epistemology: some philosophical reflections of the contemporary search for extraterrestrial intelligence', *The Monist*, January, 71, 1: 88–100.

Wilson, D. (1975) *Our Mysterious Spaceship Moon*, New York: Dell.

—— (1979) *Secrets of Our Spaceship Moon*, New York: Dell.

Wilson, E. (1995) 'New eyes for an ageing star', *New Scientist*, 14 January: 23–7.

BIBLIOGRAPHY

Wolszczan, A. and Frail, D.A. (1992) 'A planetary system around the millisecond pulsar PSR 1257 12', *Nature*, 355: 145–7.

Zey, M.G. (1994) *Seizing the Future*, New York: Simon & Schuster.

Zuckerman, B. (1985) 'Stellar evolution: motivation for mass interstellar migration', *Quarterly Journal of the Royal Astronomical Society*, 26: 56–9.

INDEX

207

Printed in the United States
by Baker & Taylor Publisher Services